THE CENTRE FOR FORTEAN ZOOLOGY
2007 YEARBOOK

Edited by
Jonathan Downes

Edited by Jonathan Downes
Cover and internal design by Mark North for CFZ Communications
Using Microsoft Word 2000, Microsoft , Publisher 2000, Adobe Photoshop CS.

First published in Great Britain by CFZ Press

CFZ Press
Myrtle Cottage
Woolfardisworthy
Bideford
North Devon
EX39 5QR

CFZ PRESS

© CFZ MMVII

ISBN: 978-1-905723-13-3

CONTENTS

Jonathan Downes
(Director, Centre for Fortean Zoology)

INTRODUCTION

After a gap of three years, it gives me great pleasure to present the ninth in the series of CFZ Yearbooks. As the name implies, these books were originally planned as annual affairs, but as history has shown, things don't always work out the way they were planned. The last volume came out in 2004, and in the intervening years a heck of a lot has changed for all of us at the Centre for Fortean Zoology.

Richard and I both lost our fathers. Richard spent much of his inheritance on our most ambitious expedition to date - a four week trek across the Gobi Desert in search of the fabled Mongolian Deathworm, and as a direct result of *my* bereavement, the CFZ has relocated to the old Downes family home in rural North Devon. Whereas for years the CFZ was based in my tiny mid-terraced house in Exeter, we now - at long last - have room to move. For many years, it was mildly embarrassing, because our visitors would look askance at our bachelor squalor, and wonder how we dared call ourselves a `Centre`. Indeed, on one memorable occasion, a visiting European TV crew said *"This ees not a Centre. This ees a small house full of books, snakes, and strange men with - 'ow do you say it? - beards!"*

They were - of course - right, but they were also very wrong. Even when we were operating out of a small red-brick housing estate in rural Exeter, we were still running the world's biggest cryptozoological research organisation, but now we have room for a proper Visitor Centre, a museum, room for our rapidly expanding collection of animals, and our even more rapidly expanding library. Whereas in 2004 there were four of us working full time at the business of running a scientific research group on a shoestring budget, there are now eight of us. Two of them are girls, and even the men don't all have beards, and wild, staring eyes any more, but we are doing the exact same thing, but doing it more efficiently than ever.

The CFZ has always been far more than the sum total of the people involved, and I truly believe that we now have an organisation which is on the verge of achieving great things, and indeed an organisation which I hope will still be running efficiently in 100 years time, when all of the current CFZ crew are dead and buried. I hope that the impetus that we have begun to achieve will become self-sustaining, and that it will continue for many years hence.

It is the diversity of people within the CFZ that are its strength, and I think that this is illustrated very well in the diversity of articles in this present volume. The range of subject matter, and writing style is dazzling, and ranges from the touching personal accounts of Oll Lewis and Lisa Dowley writing about their own experiences, and then extrapolating scientific data and theories from them, to the William Burroughsesque diatribe of Jon Hare, and the surrealchemical writing of Noella Mackenzie. We are also very proud to include the first scholarly examination of the `long necked seal` hypothesis for the sea serpent, to have been published since Oudemans in the late 19th Century.

All in all, I look at this yearbook, as I do at the CFZ as a whole, and say, in the manner of a Mafioso Don, or the proud father of a large, unruly, but loving family, and say "Guys, you done good".

Enjoy.

Jonathan Downes
(Director, Centre for Fortean Zoology)
Woolfardisworthy, North Devon
March 2007

THE BLAKEMERE
MERMAID OF MORRIDGE
AND MERMAIDS OF OTHER PLACES
(The perspective of an interested,
curious, and complete, novice)
by Lisa Dowley

De lacu in Staffordia:-

"A lake with prophetic noise doth roar,
Where beasts can ne'er be forced to venture o'er,
By hounds or men or fleeter death persued,
They'll not plunge in, but shun the hated flood.

Extract from,
History of the ancient Parish of Leek
circa 1883

As small children during the seventies, my sister and I were often taken out for lunch on Sunday afternoons, as our parents owned a Public House. One of our favourite jaunts was a drive through the Staffordshire moorlands, stopping at a quaint but remote country inn for lunch.

During our many journeys our father would tell us many a tall tale (these served two functions, methinks), and on an odd occasion we were lucky enough to glimpse the

wallabies (kangaroos as we thought), on the moors. However, the story that was always told to us, and which remains in my mind the most, is the tale of the mermaid (no pun intended, honest).

"*And this is the place where they deal with naughty girls*", our father would say to us, in a stern fatherly manner (which indicated that yet again he had grown tired of our constant bickering) as we approached the top of a steep incline.

My sister and I would turn to each other with looks of grave concern, as our behaviour between us on most days was less than satisfactory to our parents, or anybody else for that matter. "*Yes its true*", he continued, "*many, many years ago a woman was drowned here for being naughty and wicked. When she died*", he continued, his voice getting slower sterner and deeper, "*she changed into a mermaid and now waits for other naughty girls to join her!* He never *did* divulge to us the reason for her unpleas-

The Mermaid Inn

The Blakemere Pool

ant death (not that we ever asked).

Now, on the face of it this may not read as too scary, but when you are six and seven years of age respectively, it tends to make a lasting impression. Even in the hot summers of the 70s, my sister and I were loath to get out of the car and admire the view with our parents; the drop off the side of the road to us looked at least a hundred feet!

We always preferred to stay in the car; our parents probably viewed this as a success on the good behaviour and peace-and-quiet front: (mission accomplished from their point of view).

Even on warm summer afternoons, the Blakemere at Morridge - as I recall - has, and still does, retain an eerie and slightly oppressive sense about it. However, its expanding moorlands and remote bleakness still manages to be invitingly beautiful, while retaining that eternal mystical feel that totally wraps itself around you, even to this day.

Many years later, after returning to my home county of Staffordshire, the stories that our father told were still quite vivid and swirling round my head, so I decided to find out if there was any truth - once and for all - in the tales that had been told to us. Curiosity has always been at the heart of my nature, and as I had always found it odd, even as a small child, that there should be a pool so far inland that was home to a mermaid, at one of the highest points, if not the highest place in England.

Upon investigation there seem to be many mysterious, murderous and mythical tale, which may or not be merely folklore, that surrounds this bleak and remote part of the Staffordshire moorlands. Written origins of this particular mermaid folklore tale can be traced back to some 1,000 years ago.
The story transpires that this particular mermaid was once a maiden of fair beauty, and it came to pass - for reasons that are unclear - that she was persecuted, and accused of various crimes by a gentlemen, named Joshua Linnet. It is not clear whether these accusations included being a witch, or whether he may have had his amorous advances rejected.

The said Mr Linnet had this woman bound up, and thrown into the bottomless Blakemere pool. As she fought for breath and life, the woman screamed vengeance on her accuser Joshua Linnet, and that her spirit would haunt the pool from that moment hence, and swore that one day she would drag her accuser and executioner deep down beneath the dark depths of the Blakemere to his own death.

It is a recorded fact that three days later, Joshua Linnet was found face down, dead in the Blakemere pool. When his body was dragged out and turned over by the locals, to their horror, what greeted them was that what was once his face now was nothing more than tattered shreds of skin, the injuries seemingly caused by sharp claws or talons.

The mermaid of Blakemere is no bringer of good fortunes; rather she will entice any

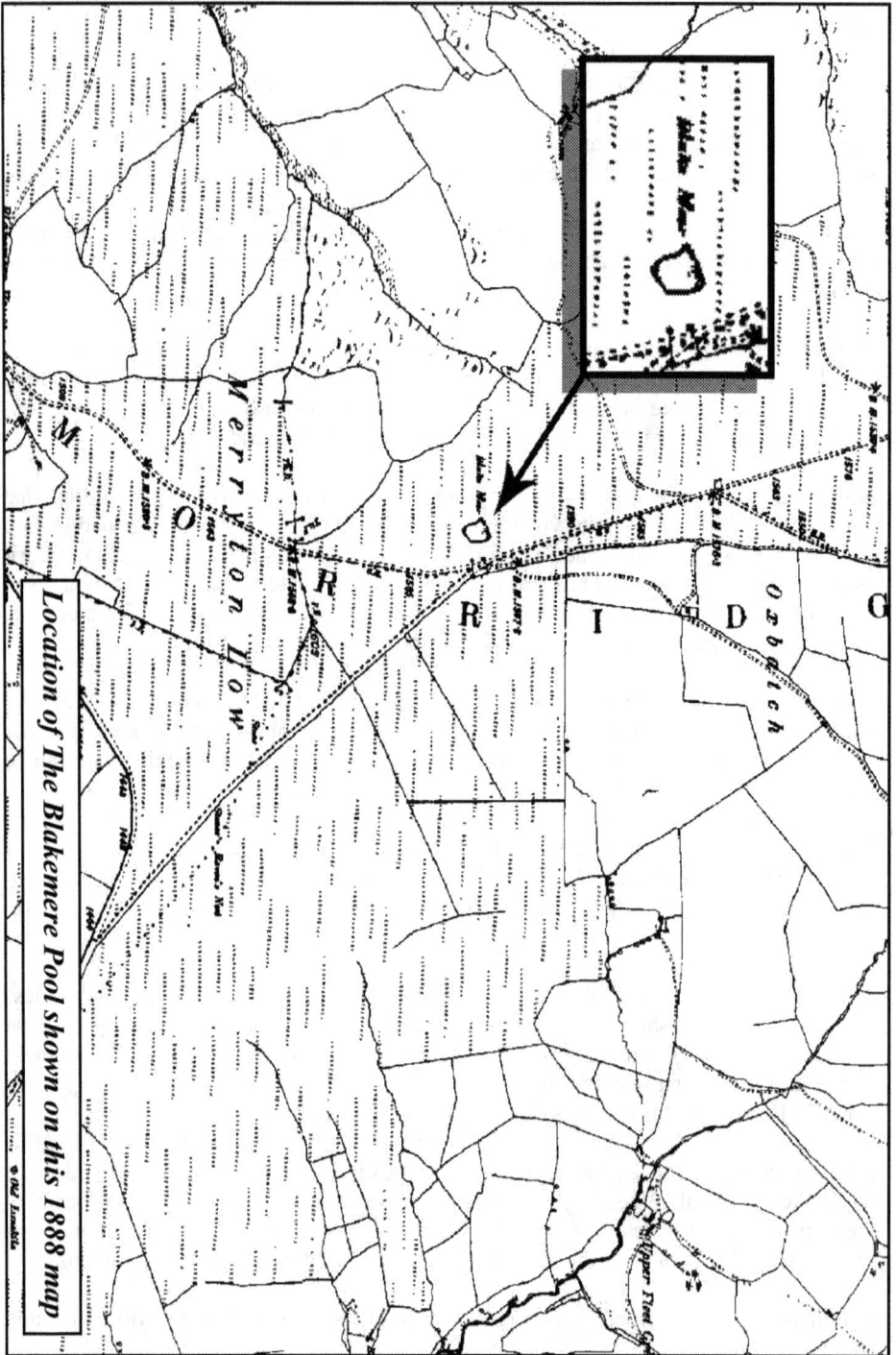

Location of The Blakemere Pool shown on this 1888 map

passer by and take them down beneath the still black waters to certain death. This is reinforced on the wall of the pub of the same name (*The Mermaid Inn,* formally Blakemere House) which is situated just below the Blakemere pool. On the wall it is written:

> *She calls on you to greet her, combing her dripping crown, and if*
> *you go to greet her, she up and drags you down.*

The Blakemere, for many centuries, continued to inspire speculation and instil fear among locals, and the belief in the mermaid persisted well into the nineteenth century. However, in the seventeenth century, a sceptic called Dr Robert Plot, a historian and naturalist of the time, was told of the legend of Blakemere and that no animal would drink from its bottomless waters or any bird fly over it. All of which he disputed in his 1686 publication *Natural History of Staffordshire.*

Towards the end of the nineteenth century work began in an attempt to drain the Blakemere. However, as the workmen attempted to drain the pool, they reported that the mermaid appeared to them and warned that if the waters of the Blakemere were removed, the whole town of Leek would be drowned. And so the men fled from the Blakemere on Morridge refusing to go back to complete the work, and so the pool remains to this day untouched and undrained.

There are a number of mermaid traditions and legends that are associated with numerous pools within the locality of the Staffordshire moorland borders, and evidence to suggest that their origins may well have been rooted in pre-Roman Celtic traditions. However, if you wish to pay the mermaid of Blakemere a visit, it is said the best time to chance seeing her is around the midnight hour. But take care, as the Blakemere pool now resides on Ministry of Defence land, so be diligent in your steps so as not to tread on any soldiers on night manoeuvres! Or you may prefer to view the Blakemere pool from the roadside from within the safety of your car.

But is there any substance that would suggest that mermaids are real and do indeed exist, albeit in other places?

Whether you believe or not, everybody knows what a mermaid is, and what they are supposed to look like. This is almost as if they are ingrained, and part of the human collective memory.

With two-thirds of the Earth's surface covered by water, and with many of these watery expanses and deep sea areas untouched and unexplored, is it possible that an entire species of creature - such as 'mermaids' - could go undetected? And is there a possibility that this creature simply has not yet been discovered and documented by man?

Professional bodies in society tend to think that a creature that is half-mammal and half-fish is nigh on impossible, seeing as these two creatures are too far apart in the context of vertebrate evolution. The general professional view is that alleged sightings of mermaids over the centuries have been no more than that of sea mammals such as manatees or sea cows, as they are often found in coastal areas but they are by no means aesthetically pleasing to the eye, unlike the many descriptions of mermaids, blaming the descriptions of beauty (albeit distorted) on the poor diet and long term malnutrition that many sailors had to endure while at sea. However this does not explain other independent sightings that have taken place from land or from non-sea travellers. Another argument in scientific favour is that many deep-sea areas have only become inhabited quite recently in geological terms that is, so the possibility of a creature from ancient mythology thriving is thought to be pretty negligible.

So on that note, does it mean that there is conclusive proof that mermaids of myth and legend do not and never have existed? Maybe not

Others put forward a theory that there is a link between `Sirens` that are mentioned in Homer's *Odyssey,* and mythological mermaids; the big issue with this theory is that where the *Odyssey* is set, the Mediterranean, dried up some 10 million years ago, and only refilled relatively recently. Even Homer makes reference to the sea being as dry as a desert, ergo ancient stories from the `Med` containing such creatures that have endured from ancient times could not be possible. Could they?

Having said that, mermaid stories and legends are not just based around the Mediterranean, but are culturally universal, and can be found in legends all around the world. They are important not from just a cultural aspect, but often also have a great religious significance; and they *all* display striking similarities. Many of the references to mermaids are reported from bodies of water where there are no known species of sirenians (manatees, dugong etc) at that given time.

The first recorded reference to a part-fish part human (mermaid/merman) is the ancient Babylonian belief in the God Oannes, Lord of the Waters.

Oannes

SIRENIA

An order of fully aquatic, herbivorous mammals that inhabit rivers, estuaries, coastal marine waters, swamps, and marine wetlands. The order evolved during the Eocene epoch, more than 50 million years ago. **Sirenians**, including manatees, the dugong,, and the extinct Steller's sea cow, have major aquatic adaptations: forelimbs have modified into arms used for steering, the tail has modified into a paddle used for propulsion, hindlimbs (legs) are but two small remnant bones floating deep in the muscle. They appear fat, but are fusiform, hydrodynamic, and highly muscular. Their skulls are highly modified for taking breaths of air at the water's surface and dentition is greatly reduced. They have only two teats, located under their forelimbs, similar to elephants.

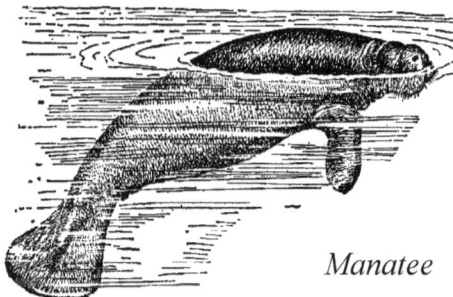

Manatee

Dugong

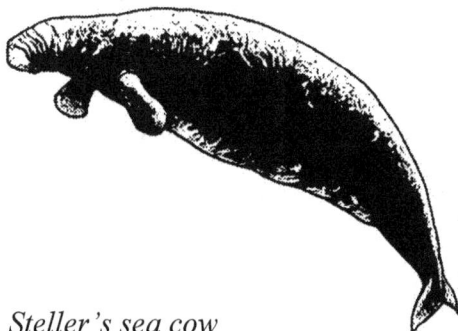

Steller's sea cow

- In Germany there is the *Merminni* or *Meerfrau,*

- Iceland has the *Hagufa* or *Margygr*

- The Danish use the term *Hafmand* or *Maremind,*

- The Irish - where a great deal of the British mythology potentially origi nates - use the name *Merrow* or *Merruach.*

- More recently there is a curious cultural transformation occurring in certain parts of Africa, whereby the image of the mermaid (known as *Mama Wata*) is replacing the originally envisioned water spirits and creatures such as snakes and other reptiles.

No matter which cultural strain of the mermaid myth you choose to follow, most - if not all - display similarities in their accounts of physical descriptions of these creatures. *If* we are prepared to cast aside the Hollywoodesque and media hype of the 'scale clad golden haired beauty' a seemingly different picture emerges.

- Mermaids have been reported as having long flowing dark hair, some re- ports - especially sightings off the coast of the Isle of Man - have made refer- ence to a reddish colour; but, in the most part, the hair is long and dark (possibly black) in colour.

- The ears have been noted as being rather long and semi-pointed. This may be an adaptation to being in the water. The eyes have appeared dark, and more of a large oval than round shape, again this could be an adaptation to the wa- ter.

- The nose has been described as depressed or flat and resembling an Apeoid/ Mongoloid species of human.

- From various accounts, the upper half of the female mermaid body is reputed to resemble that of a human woman, with large perpendicular breasts. The skin tone has been described as being from a fish-belly white, ranging through to a light grey/blue white.

- The lower half of the mermaid has been described as having a fish-like tail, but appearing more mammal-like; reminiscent of that of a dolphin, seal, or manatee. In most descriptions there is no mention of any scales on this tail appendage at all. The colour of the tail seems to vary, from a reddish-brown, dark brown to a speckled type pattern of varying shades of green/blue/grey.

- The male of the species has been described as quite similar (minus the perpendicular breasts!) and of a more unpleasing aspect to look upon. The over-all length of the creature is reported to be between no more that 4 to 5 feet in total length.

Yet again curiosity getting the better of me, I began to wonder how can differing societies across the globe have such similar descriptions of these merfolk. More to the point, how can these myths and legends share so many important strikingly similar features and themes? These themes seem to be a constant through varying time spans, and possibly our subconscious.

Then quite by chance, I happened across a practically unheard of and little publicised - let alone discussed - theory, which to a complete curious novice such as myself seemed quite plausible and makes a great deal of sense compared to - say - algebra (I never understood swapping numbers for letters!).

In 1960 Sir Alister Hardy published an article in the *New Scientist*. In this paper he put forward for consideration a hypothesis that suggested that man has passed through an aquatic phase, suggesting that this was the reason for the many differences that separate us from apes. This theory became known as the 'Aquatic Ape Theory'.

This `Aquatic Ape Theory` is situated in a time of great drought not agreeable to our ape ancestors, To survive - he theorised - these ancestors needed water close to hand in order to thrive (taking into account the fossils before and the fossils after this era, there are almost none to be found during this time) and it is thought that this period in environmental time had a profound effect on our evolution. His original theory suggests that the physical changes that took place are due to the climatic change and the prolonged exposure to water.

These ape ancestors were used to climbing trees in order to evade predators, and to search for food, but, due to the climatic change, most of the trees were gone, so - in order to survive - they had to adapt to their environment, and were forced to search the shoreline for food, and also use the water for protection. He further postulated that from this shore-line scavenging they began wading further and further into the water, until they were able to swim for a time, intermittently resting with their feet on the bottom with their head out of the water. Indeed, he speculated that this could have given support to our species having an erect and bipedal posture, as the water may well have served as a support mechanism for our ape ancestors' weight.

There are many evolutionary changes that Hardy attributes to this aquatic ape phase in comparing ourselves to other aquatic mammals.

Seals, whales, dolphins, manatees, and hippos, were all once land animals which moved into water, and over evolutionary time this exposure caused various body adaptations. In the case of seals, whales etc., fins are the remnants of limbs. The water also - over time - caused these animals to lose most of their fur: as fur inhibits swimming.

THE
AQUATIC APE
HYPOTHESIS

Elaine M____n

AUTHOR O_
The Descent of

CREDIBL_
HUMAN

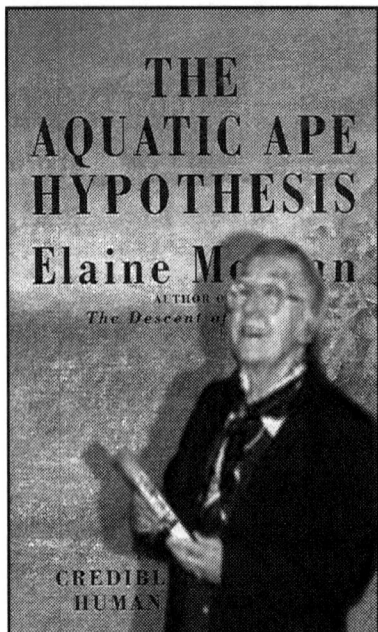

Elaine Morgan (born 1920) is a Welsh feminist writer, best known for her television work, including screen writing most of the episodes of *Dr. Finlay's Casebook.* She is also the author of several books about the aquatic ape theory, among them *The Descent of Woman*, *The Aquatic Ape*, *The Scars of Evolution*, *The Descent of the Child* and, her latest, *The Aquatic Ape Hypothesis.* She also authored *Falling Apart* and *Pinker's List.* Morgan is generally described as more of a popularizer of science than a scientist.

In 2003 she started to write a weekly column for the Welsh National Daily, *The Western Mail.*

She was awarded an honorary D.Litt. by Glamorgan University in December 2006.

From Wikipedia, the free encyclopedia

Chararacteristics	Humans	Apes	Savannah	Aquatics
Habitual Bipedalism	Yes	-	-	-
Loss of body hair	Yes	-	Yes	Yes
Skin-bonded fat deposits	Yes	-	-	Yes
Ventro-ventral copulation	Yes	Yes	-	Yes
Dimunition of apocrine glands	Yes	-	-	Yes
Hymen	Yes	-	-	Yes
Enlarged sebaceous glands	Yes	-	-	Yes
Psychic tears	Yes	-	-	Yes
Loss of vibrissae	Yes	-	-	Yes
Volitional breath control	Yes	-	-	Yes
Eccrine thermoregulation	Yes	-	-	Yes
Descended larynx	Yes	-	-	Yes

Thus, over time, our ape ancestors would also have lost a great amount of their fur, but possibly retaining long head hair so that infant apes had something to cling to.

Another possible reason that is offered for losing most of our hair coverage, is that fur does very little to keep an animal warm underwater, and wet long fur serves no purpose but to keep an individual cold and slow them down. However, a layer of subcutaneous fat, like the blubber of whales and other aquatic mammals, keeps the internal organs warm. Humans are the only primates who have such a layer of fat; could this be another remnant of our aquatic ape phase?

Furthermore, what body hair that we still possess lies and grows in the same direction as it would if water were passing over it, say as in swimming. Another change that separates us, is that when aquatic mammals enter the water, their metabolism decreases, which serves to enable them to stay underwater for longer periods of time. This can also be said of humans, albeit ever so slightly - however, it still nevertheless slows down. Could this also be interpreted as a remnant of our time as aquatic apes?

All of Hardy's theories above (and more), were addressed and expanded upon in a later released book in 1972 by Elaine Morgan titled *The Aquatic Ape, A Theory of Human Evolution*. Then twenty years later she re-addresses the subject in her follow-up, *The Aquatic Ape Hypothesis*.

Morgan has no professional scientific background (biological or anthropological) as such. Her roots lie in writing. This is rather refreshing, as you don't find yourself getting lost in academiaspeak. She approaches the subject in a very balanced, plain speaking, common sense manner, with a convincing collection of evidence which gives credence to our ape ancestors having gone through a semi-aquatic phase, incorporating various other theories such as the savannah, and neoteny theory. But before you totally dismiss her writing and views as having no academic merit, may I point out that neither did Darwin! (His schooling background was in Theology)

I am not suggesting that this theory is absolute, but it *does* make for a possible alternative, coherent, explanation of human physiological adaptations and features that fit into what is at present an evolutionary gaping void, which scientists seem hell bent on trying to fill by inventing elaborate theories that only make sense to themselves.

The `Aquatic Ape Theory` is an under-discussed theory that merits some long overdue consideration. Indeed: obtain the book, read it at your leisure, and draw your own conclusions.

However, I am quite aware that there will be individuals out there screaming: 'Where's the proof? - Where's the evidence for these aquatic apes? Or indeed for mermaids?'

Well one plausible retort could be: creatures that die in or near the water are more than likely to be washed out to sea, where they will be quite likely be eaten by other sea creatures or decompose or dissolve in the seawater. If fossil evidence of these aquatic ape creatures *were* to be found, then maybe the best place to find them would indeed be the sea bed or beneath the ocean floor? As far as I know, we have not become technologically advanced enough to attempt these sort of archaeological submersive excavations, but I am quite sure that in time that we will. And who knows what evolutionary secrets we will find beneath the sea bed?

To give more credence to there being an unidentified aquatic ape creature, an interesting news report was obtained by the CFZ, in March 2006, regarding a mysterious amphibious creature, which was sighted in the Caspian Sea.

For the last two years, residents of coastal areas around the southern and south-western Caspian Sea have been reporting some unknown amphibious creatures resembling a human being. In March 2006 an eyewitness account from the crew of the *Baku*, an Azeri trawler was published by Iranian newspaper Zindag:

"That creature was swimming parallel course near the boat for a long time". Said Gafar Gasanof, captain of the ship. "At the beginning we thought it was a big fish, but then we spotted hair on the head of the monster and his fins looked pretty strange.... The front part of his body was equipped with arms."

Back in Azerbaijan, nobody took his story seriously. It sounded ridiculous; many thought that the guy must have been drinking while on board. On the contrary, shortly after the publication of his interview, the offices of the Iranian paper were flooded with numerous letters from the readers, who claimed that the story was yet another piece of evidence proving the existence of the so-called "man of the sea". The readers pointed out that many fishermen had repeatedly seen the strange creature at sea, *and* on shore, after the seabed volcanoes in the area of Babolsera had come to life in February, and offshore oil operations had intensified in the Caspian.

Iranians dubbed the creature *Runan-shah* or "the master of the sea and rivers". The name is partly based on stories about shoals of fish accompanying the creature at sea. Other stories refer to the waters that would turn crystal clear, and stay that way for two or three days after the creature was seen swimming in those areas. Fishermen claim that fishes that stay alive for a while in the net, can feel the creature coming out of the deep blue sea. Fishes were reported producing barely-heard gurgling sounds, as the monster came near. It was thought to answer the call of the catch by making similar throaty sounds.

Some researchers believe that there is no smoke without fire, and that the stories circulating in Iran *could* be true. Besides, in May 2006 *Runan-Shah* was seen by Azeri fishermen living in the villages located between the cities of Astara and Lenkoran. It has been suggested that the creature is not alone, and that there is a family of underwater humans.

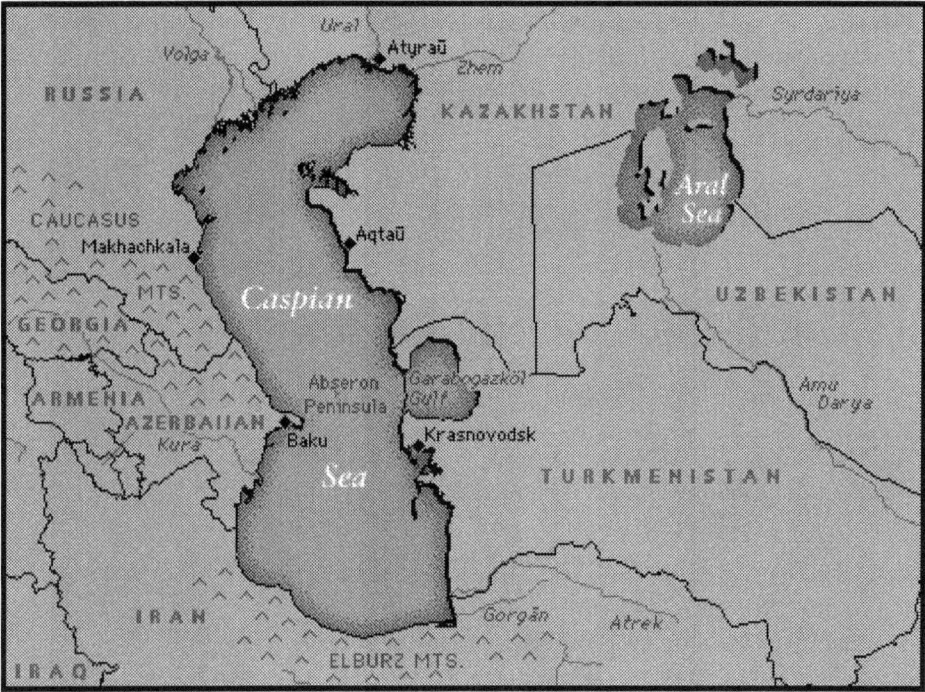

The Caspian *Runan-shah* is not the *only* species of underwater humans on record. In the early part of the twentieth century a human-like creature was reported in Karelia in 1928. The creature was repeatedly seen in the lake of Vedlozero by local residents. A group of researchers from the Petrozavodsk University arrived to investigate the case on location.

However, the findings from this research trip were classified and even more curiously members of the research party eventually perished/disappeared in the Gulags. Going on the latest, and current reports in the media, Iranians have already started their research of the Caspian phenomenon.

The seas and oceans are great and vast places. The thought of such a place harbouring such creatures as mermaids or indeed aquatic apes could well be plausible, and only time will tell.

This was demonstrated in the latter part of the twentieth century many unknown and indeed thought to be extinct species of animal were discovered. In 1983, for example, the Prudes Bay Killer whale was discovered. This is by no means a *small* creature. Who knows what the twenty-first century holds in store for us? It is a very difficult task to study such elusive and little known creatures alive *or* dead, as even creatures in relatively small bodies of water have often proved almost impossible to observe or capture.

It is however, an ancient belief that `thoughts` are `things`, and you only need to think of something strongly enough for it to indeed become real, if something can exist with such prominence in the human psyche, the chances are that it is real.

What do *you* think?

SOURCES

The Mermaid Inn, Morridge Top, Leek, Staffordshire
History of the Ancient Parish of Leek , John Sleigh Circa 1883
The Folklore of Staffordshire J. Raven 1978
The Aquatic Ape, A Theory of Human Evolution Elaine Morgan 1982
ISBN – 0 285 62509 8
Modern Mysteries of the World, 'Strange events of the 20th Century'
Janet & Colin Boad 1990 ISBN 0-586-21028-8

WEBSITES

Mermaids and Mermen, LoveToknow 1911 Online Encyclopaedia
2003, 2004 LoveToknow
http://2.1911encyclopedia.org/M/ME/MERMAIDS_AND_MERMEN

Water Spirits and Mermaids: The Copperbelt Case
Brian Siegel 2000
http://www.ecu.edu/african/sersas/Siegal400.htm

The Centre for Fortean Zoology
www.cfz.org.uk

MAN-TIGERS

by Jon Hare

Monsters: *from the Latin monstrum': 'that which is shown forth or revealed'.*

Monsters: *identified by Isadore of Seville as monstrations (monere) or warnings (monare) of divine will.*

168

George Cram, 1897, Sumatra. Printed colour, about 12 1/2 by 8 3/4 inches.
Shows the districts, cities, towns and landforms.

It begins with a fire. Around it, an audience, and a story. *"There are men like us, but not like us,"* this story goes. Rings of bright young eyes throughout Sumatra hear this tale, surrounded by the green forever of the jungle, fireflies, the coughing of tigers. *"In the mountains, in the shadow of the volcanoes, in the jungles, outside us and amongst us. Cindaku. Ngelmu-Gadongan. The Jadi-Jadian. Tigers who are men; men who Are tigers."*

This story is told all over Sumatra. It was told last night in Padang and it will be told tomorrow night in Sungai Penuh. It was told on the 28th of June, in 2003, while I was spending my first night in my first jungle, dozing fitfully in a sleeping bag and feeling something sudden and hot and silent and heavy brush past the side of the tent on padded feet, frozen and helpless as it crept up to lay its breathing form down next to me in the darkness, sliding a paw over me and filling me with a warm peacefulness that blew me out like a candle on the hard ground by the side of the lake. I woke with a start, and early. Mentioned the experience to our guide, Sahar, who freaked out as if I'd punched him in his third eye.

My guide, it later emerged, was a weretiger. Later, I learned of a legendary village of weretigers rumoured to exist somewhere in Sumatra's mountainous interior.

Later still, he took me there.

But this is one story among many. For many reasons, tigers are important. A dukun can contact them and call them. A warrior can draw on their power. They carry a shaman's soul through the jungle on whisper-quiet feet. They teach the warrior everything he knows. More than animals: the crossing point of man and nature in the trackless forest, the keeper and gateway to forgotten ancestries. A creature to respect. Once upon a time, we called our spirits `Good Folk`, `Kindly Ones`. Sumatrans say that tigers are 'polite' and leave it at that.

So, was it politeness that it came to me, this spirit-tiger? Did it come to welcome me to the jungle? Twenty years ago, to a boy lying on his bedroom floor, face-down in an encyclopaedia of animals, Sumatra defined jungle. It was ultimate jungle. The Amazon was piranhas and lost cities; Africa, Tarzan, gorillas, lions. But Indonesia was where dragons lived and Sumatra ...

Sumatra was man-eating flowers, giant snakes, bright tigers. It was *Where the Wild Things Are:* the forest at the bottom of a child's mind where boundaries of self and environment, present and past, blur; the place anthropomorphism is born, where animals and men play together, indistinguishable.

But there's little jungle left. Deforestation began as a coastal malaise, a few rectangular patches on a blank map, but after twenty years of metastitization and colonisation the island has been coloured in. The forest is archipelago where once it was continent; puddles not ocean; peaks shrinking from the encroaching sea of plantations, shanty

Our guide in Sumatra, Sahar

town, civilisation, for it is only in the mountains that the chainsaws cannot go.

The island of Sumatra is staunchly Muslim. Alarm clocks and cockerels are beaten to the punch every morning by crackling amplifiers and the dawn call to prayer. Iron-mongers selling household minarets are common as corner pubs. But don't be fooled: it's a thin paint, a cover-up job of buttoned-down religiosity, suspicious as the mirror-smooth paintwork on a car-dealer's forecourt. Something must be up, and it is, for underneath is a bubbling animist core that goes right back to the the jungle, sweating through the orthodox surface despite all that religious deodorant, a living tradition of spirits and the supernatural.

Tigers are a very, very important animal in Sumatra, particularly to the people living closest to the jungle. The Sumatran tiger is ... beautiful. The smallest surviving sub-species and the most brilliantly coloured, respected but not feared. The tiger is 'polite', they say, but as a magical animal, requires politeness in return, although man-eating tigers are uncommon in Sumatra. However, when walking in the jungle, great care is taken not to 'upset' the local tigers. Nudity, greediness, all invite trouble.

The shrinking jungle is dotted with villages and every village has its doctor, its DU-KUN. Their crystallised elephant sperm will cure your cold if you care to try it. There are an awful lot of fake Dukun, particularly in Java. 'Photocopied doctors', the locals call them.

But the real ones are also shaman, and shaman the world over are considered to have the ability to call and control animals. In Sumatra, A DUKUN HARIMAU is a shaman with a close relationship with the tiger. He may go to sleep having burned special in-cense, then go to the forest and there walk with a tiger.

Three years ago, on an expedition looking for ape man in the Sumatran jungles, I met British journalist Debbie Martyr, head of the Tiger conservation team in Gunung Ker-inci national park. Debbie knew the shaman well. The year before, when tourists be-came lost in the jungle, she asked a Tiger Shaman to look for them. He stayed in the village that night ... but his tiger went looking. The next day, a search and rescue team found the tourists safe and well, surrounded by fresh tiger prints.

(As Dukun have the power to call animals, I asked if they could call my Sumatran ape man. *"It's useless,"* she said. *"I tried once and they insisted they could find him, but after we'd trekked into the jungle for a bit, a tiger appeared from nowhere, growling and angry! They didn't want me to be disappointed, so they'd called a tiger instead. I shouted, 'Who called the ancestor spirit??' 'Not me!', they replied. 'Why is he angry? I heard you call him!' 'Well ... he had to come a long way and we didn't warn him in advance'!"*)

• Weretigers

CINDAKU (pronounced 'Tjindaku') are human-appearing monsters considered to eat people and naughty children in Sumatra. The word is often used to mean weretigers in Sumatra, especially 'harimau cindaku', although according to older sources, the proper term for a magic tiger or weretiger among the Minangkabau people would be 'URANG JADI-JADIAN', probably a corruption of the phrase 'worsening man'.

The base of the tiger conservation team is called Sungai Penuh. It nestles below the slopes of Mount Kerinci and Sumatrans consider it to be the absolute epicentre of weretiger activity in the country. *"People say, 'Don't go to Sungai Penuh! The locals are all cindaku!'"*, although if you say this to people from Sungai Penuh, you'll get punched!

The **Sumatran tiger** (*Panthera tigris sumatrae*) is found only on the Indonesian island of Sumatra.

The wild population is estimated at between 400 and 500 animals, occurring predominantly in the island's national parks. Recent genetic testing has revealed the presence of unique genetic markers, indicating that it may develop into a separate species, if it is not made extinct.

This has led to suggestions that the Sumatran tiger should have greater priority for conservation than any other subspecies. Habitat destruction is the main threat to the existing tiger population (logging continues even in the supposedly protected national parks), but 66 tigers were recorded as being shot and killed between 1998 and 2000—nearly 20% of the total population.

Corinna James

William Blake - The Tyger

Tyger, Tyger, burning bright
In the forests of the night,
What immortal hand or eye
Could frame thy fearful symmetry?

In what distant deeps or skies
Burnt the fire of thine eyes?
On what wings dare he aspire?
What the hand dare seize the fire?

In that region, a 'cindaku' is not a man who turns into a tiger. It is not a tiger who turns into a man. It's both, and neither: a tiger spirit.

The legend of the NGELMU-GADONGAN, the man-tigers, is thought to originate from the West Sumatran district of Pelambang but it is widespread among the jungle communities. The Sumatran *ngelmu-gadongan* is different than the Javanese version, a cursed being like a Western werewolf, or the Argentine IRUNAUTURUNCU from the jungles of South America, a sorcerer possessing the mind of a man but the power of a tiger, through a pact with the devil (ZUPAY).

Sumatrans say that a group of human beings exist with the ability to turn into tigers.

They are seemingly normal in every other way but they can be identified by a single physical peculiarity: they all lack the channel in the upper-lip. Most of the time the cindaku stay in human form and live just like any other people, but at certain times of the year they abandon their homes and head off to hunt. When a hunting *cindaku* arrives at a neighbouring village, he will stay in human form, entreating the villagers to allow him to stay the night. If the locals are not wary and do not notice that he lacks the channel in the upper lip, the tiger will transform in the night and devour them all: in the morning, all that will be found are bones. The *cindaku* will have melted back into the jungle.

According to Pelembang legend, there is a village in the mountainous Dempo region inhabited exclusively by man-tigers. It is entirely possible that this is Sungai Penuh, centre of the tiger conservation project.

• Martial arts and the tiger

Pencak Silat is the national martial art of Indonesia. *Harimau Pencak Silat* comes from the Painan region of West Sumatra, although its influence can be found in every type of *silat* throughout the rest of Indonesia. More than any other martial art, more than the clawed hands of Chinese Shantung Black Tiger or the ferocious aggression of Burmese Tiger Bando, *Harimau persilat* try and embody tigers, truly become them. Their power is such that entire styles have been built around fighting them.

I trained in Sumatran *Harimau silat* for two years. Getting closer to the tiger is one of the reasons I went to the jungle. The energy is feline and feminine.

Designed for jungle fighting, they fight on all fours, a way of moving unseen, stealthy, tigerlike in the undergrowth but also a way of building an inner tiger, not in imitation, but something deeper than that. A pathological process, changing body image, feeling a tail waving behind you like a phantom limb, feeling yourself with four hands, four paws, and feeling the ground under them. A form of possession.

Through intense practice and visualisation techniques, the *persilat* will begin to see his hands as paws. He will feel his tail swishing behind him and balancing him as he moves. Sometimes, a 'phantom tail' may develop as body-image begins to shift ... like a phantom limb, sensation and discomfort may occur if it is inadvertently 'walked through', for instance. (Don't step on his tail!) Further, a powerful *harimau* man will begin to be seen by his enemies as an actual tiger. The imitation will fall away and transformation will occur. Real or unreal ... who's to say? His enemies will be torn apart as if by the claws of a giant cat.

The top masters of *Silat* are called PENDEKAR, a word meaning 'spiritual champion'. They are very rare now and many believe that there are only a few left. The *pendekar* is not just a physical warrior. He has magical powers and in the *Harimau Silat*, the *silat* of the tiger, he also has a special relationship with the tiger. The tiger is his Guru, his teacher and master, and it is to the tiger he must go to be tested.

My jungle guide Sahar's village has a very special relationship with the tiger. He told us that sometimes, when a *persilat* (a practitioner of Silat) becomes very good and is walking in the jungle, a tiger will 'test' him. The tiger will appear, and will leap at him, so the *persilat* must evade and roll away. The tiger will land and leap again; again, he must evade. This can go on for ages.

When Debbie first moved to Indonesia, her standing among the locals, as a single woman, was a little problematic. Then something extraordinary happened.

Debbie was walking jungle trails with her guides and spotted strange hairs on a bush. To the astonishment of her guides, as she crouched to look at them, a full grown tiger leaped suddenly from the jungle and sailed over her head, vanishing soundlessly into the trees as quickly as it had emerged. In a single moment, this act, which seemed to confirm the *Silat* legend, allowed her to gain the trust of the villagers. She'd been tested and found worthy by the ultimate authority in the jungle. After that, who would question her?

In the West, a weretiger is a human with fur and claws, or a man with a tiger's head, but the East is less literal. In Sumatra, adrift from a binary Western worldview, a place where black is black and white is white, where a man and a woman are North and South and animals and men separated by uncrossable barriers, the locals find the idea of a transformation that's physical yet not-physical to be perfectly acceptable. They're comfortable with something inhabiting two states simultaneously. A man can be both a man and a tiger, simultaneously. He can look like a man, yet be a tiger inside.

Monsters fascinate and scare us in equal measure. The late French deconstructionist, Jacques Derrida, was fond of pointing out the lessons the uncanny could teach us about our worldview. Analysing what we're scared of, he said, the monsters that bubble up out of time and culture, shows us things our cultures are scared of, too.

Monsters destabilise our Western ontological bedrock, the normative conceptions

Harimau silat is big business - you don't get a sponsor like Panasonic by being a small-town pursuit, followed by a few hundred locals around which we structure our everyday lives, which are largely based around the creation of a network of stable opposites. Mind and body, real and unreal, nature and culture, animal and man. Our comfortable assumptions are balanced precariously on the walls we erect between those states. Monsters are frightening not only because they're going to sink their fangs into our throats, but also because they break through those walls, cutting the ontological ground out from under us, destabilising our comfortable philosophical shorthand. The unearthly, therefore, can inspire true horror, but also, perhaps, has the power to liberate us from the banal.

Zombies are both alive and dead simultaneously and it's this which gives them their uncanny power. You either have to make them alive or dead, 'decide' them one way or the other, otherwise your worldview has a huge problem on your hands! They'll eat your brains and scramble you! Vampires are also both alive and dead, werewolves are simultaneously human and animal, robots are inorganic yet exhibit the characteristics of the organic. This is a liminal place, a form of philosophical twilight: *'entre le chien et loup',* as the French put it.

Sahar's ancestors were weretigers. His people live in a small settlement surrounded by miles of trackless jungle until only a dozen years ago.

John

"One of my ancestors, the founder of my people, lived to be 150 years old and then died, but he became a tiger. Those of his bloodline have this sympathy, this DARAN BURUN, this spiritual mark. It's in our blood, it's unclean, dirty blood and it's very dangerous for us to go into the forest, so we must be cleansed. We will ask the tiger, who is also our ancestor spirit, to look after his children, even if we have this mark that means the tiger can take us. We will say: 'Please look after your grandchild and remember you were a human before you became a tiger and do not eat me'. The cleansing is done by the village Dukun and involves TARI ASIK spirit dancing to call a NENEK possessing spirit. Once the Nenek enters me, I am not Sahar anymore, but the tiger. I growl like a tiger and will search for and attack the person who has the tainted blood. If I pick out Jhon, my brother, he will be very frightened because I will pounce and sink my teeth into him, biting to remove the taint. It can be very bloody, but afterwards, Jhon can enter the forest safely. The ancestor-tiger will protect him against the living tigers in the forest."

me

"Sometimes, there's blood everywhere. When the tiger enters them, they _are_ a tiger. The hair stands up on the back of your neck. But it's not touristy. They believe it's impolite to let a Nenek enter them if anyone outside the village is watching. It's not a public spectacle. When they dance the Tari Asik for an audience, you can sometimes see one of them struggling as a spirit tries to enter them and they try to fight it back, the whole village will notice and try to stop the possession taking place before they be-

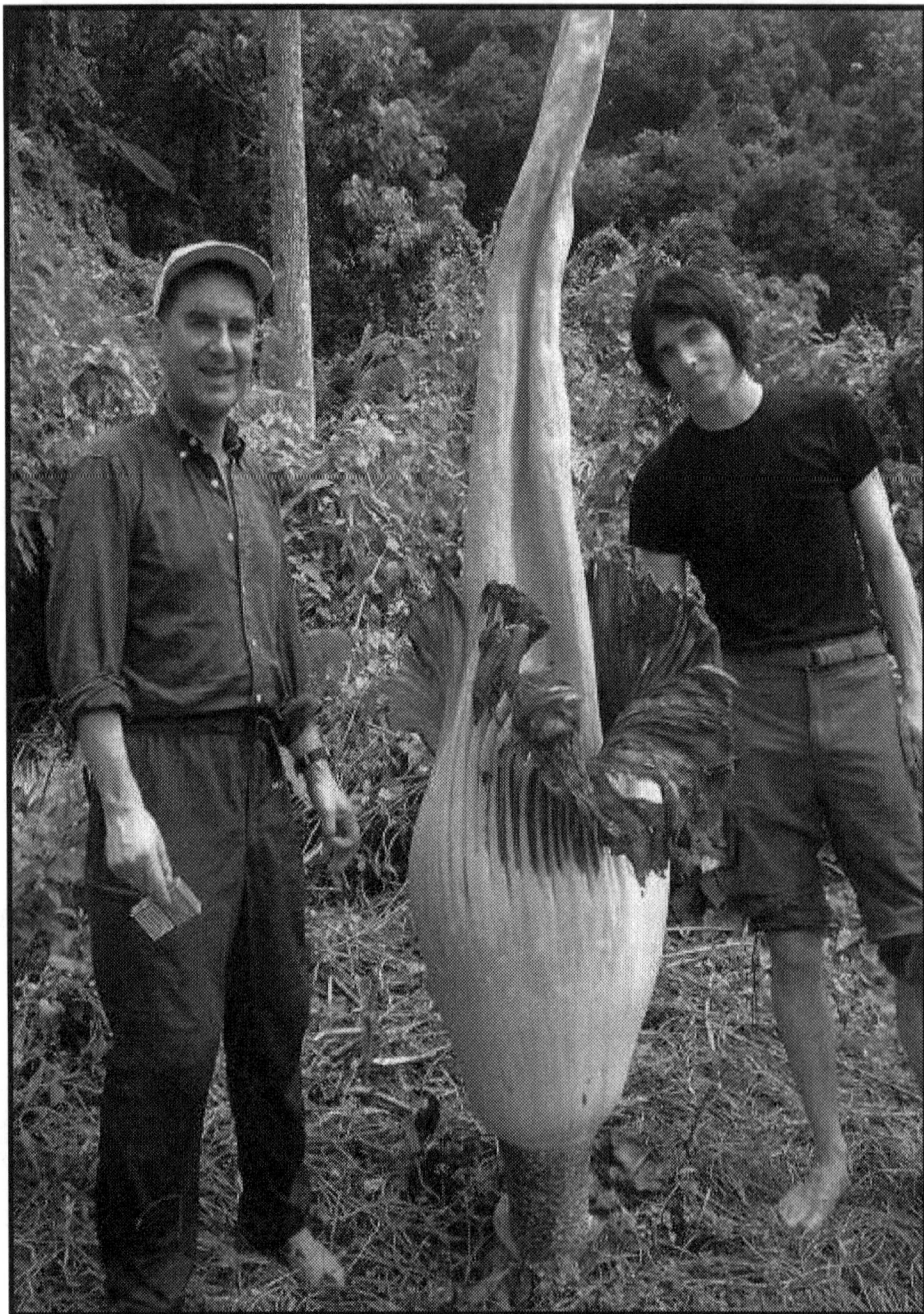

Jon Hare, (L), and Dr Chris Clark in Sumatra 2004.
They are standing by a titan arum - the world's biggest flower

Jon Hare at Mount Kerinci (2003). The tiger statue was erected because it is the sacred animal of the people in that area

The 2004 CFZ Sumatra expedition: L-R `John` (guide), Dr Chris Clark, Jon Hare, Richard Freeman

come a tiger and start mauling people

me

I never saw this, but Debbie did, and she may have been the only westerner to actually see a real weretiger.

• A quick brainstorm:

Weretiger/lycanthropy ... deconstructionism is interesting, like Jacques Derrida's 'both/ and' principle - monsters destablise our Western philosophical bedrock which has created a network of stable opposites.

Monsters are frightening 'cos they cut the ontological ground out from under us, destablise our philosophical shorthand, our tendency to create dialectic, to split things into opposites, to categorise, zombies (alive and dead simultaneously), weretiger (man and animal simultaneously), hermaphrodites (man and woman simultaneously), robots (organic and inorganic simultaneously) ... Blacksmiths making a sword ... fire & water, earth & air, the Kris, metal from the stars ... alchemy where opposites fight, dissolve, recombine (solve et coagula - irreconcilable opposite are 'solved' through a mystical union) to form a higher union (Hegel's historical progress; Marxism), creation through destruction, taking something apart (massa confusa) to make something pure (prima materia) (filtering), turning SHIT into GOLD (late stage capitalism - fuckloads of filth being peddled for big bucks, but has Rupert Murdock discovered the Opus Magnum? Has he bollocks - Spagyric is the art of uniting through dividing ... uniting all of our money by dividing us - pretty good summary of the tabloids if you ask me but the Sun would hang a weretiger let alone a hermaphroditic rebis faggot); "myth is already enlightenment; and enlightenment reverts to mythology"; Jung's work on alchemy & psychology, the dissolving of hardened positions. Sulphur & Mercury ... solve et coagula is the motto branded on the forearms of the devil, according to Levi!

That's about it.

THE MYSTERY MENAGERIE OF GLAMORGAN

by
Oll Lewis

I
t happened as dusk set in one wet but unseasonably humid winter's night. In the picturesque Victorian seaside town of Penarth, a man named Evan Davis was driving down the tree-lined avenue of Bridgeman Road. As Evan's car crept towards Alexandria Park, out from among the trees and fallen leaves darted an unusual creature. On first glance Evan thought that the animal was a cat, but something didn't quite scan; if it *was* a cat it was certainly a very large one; it had long black-grey grizzled fur, and a long bushy tail with two distinctive white rings. Evan was in little doubt he had seen something out of the ordinary, and wrote to the local paper in an effort to unearth more information about it. Did anyone have any idea what it may be? Had anyone else seen it?

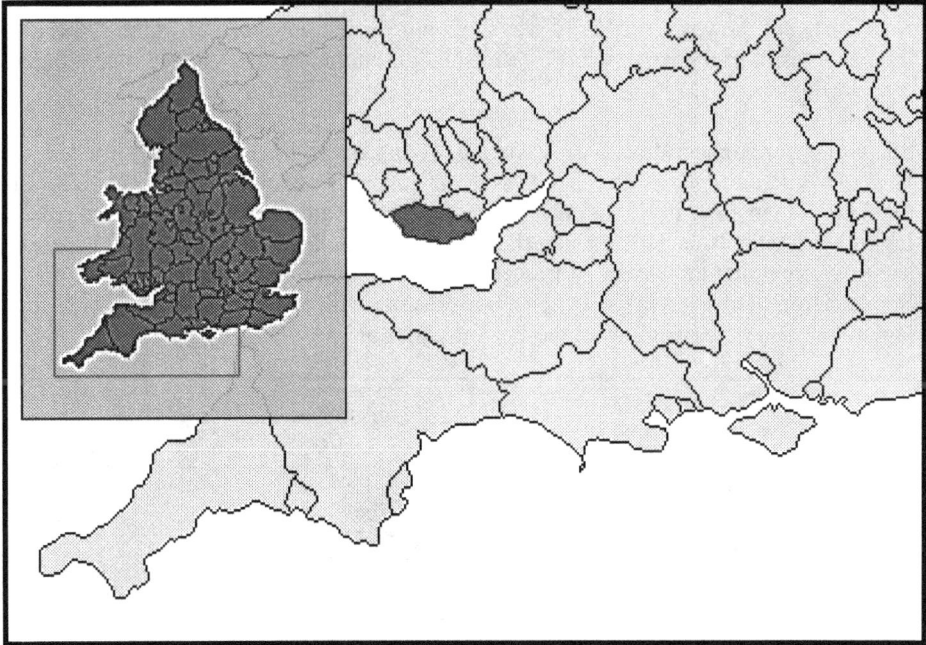

The reporters at the *Penarth Times* were certainly intrigued enough to print Evan's letter. After all this is Penarth; it's on the outskirts of Europe's fastest growing capital city, the boom town of Cardiff. Things like that aren't seen here... or are they?

Cardiff and Glamorgan have had more than their fair share of strange beasties hopping around the place - in the case of the Cardiff kangaroo a few years ago, quite literally.

The kangaroo, after a few days of being spotted loitering in Llanishen's gardens in the north of Cardiff, was never seen again, and from video footage, the local R.S.P.C.A. concluded that it was probably a fox with mange. Cardiff and its environs has been the reputed home to more than one genuine cryptid though. In one local myth dating back at least to medieval times, the river Taff has been the home of a giant serpent. This serpent was said to have made its home near a bend in the river, and caused whirlpools to appear whenever it coiled or uncoiled its sleek, slender form. What happened to the monster is unrecorded and since the Taff is now, thanks to the Cardiff bay barrage, non-tidal, it's very rare that one might see a whirlpool.

Another cryptid reputed by legend to haunt the area was the Gwiber, a winged serpent that was in some cases benevolent but mostly took delight in terrorising the local townsfolk, only to be killed by a local shepherd or workman who tired of his town's constant persecution. Gwibers were more common in mid Wales, but Glamorgan's Gwibers were associated with the woods of Penllyn, 14 miles north-west of Penarth and recorded sightings took place into the 19th century.

A mystery cat, however, certainly was new, but there were more oddities to come to light as a result of the paper's call for information on the mystery cat and other strange creatures in Penarth.

Something far stranger emerged. A woman, asking not to be named, went into the offices of the *Penarth Times* to relate the tale of a most unusual, and sinister garden visitor. The woman, who lived very near to the vast graveyard of Penarth, had been pottering in her garden when she felt like she was being watched. Looking over to her fence she saw something that would scare her from even venturing into her garden for days. Spread across her fence she saw a bird like creature with a wingspan of several metres, larger than that of the window of the Penarth times office.

Penllyn Castle - the haunt of a gwiber as recently as the 19th Century

THE WINGED SNAKES OF PENLLYNE CASTLE

The woods around Penllyne Castle, had a reputation for being frequented by winged serpents, and these were the terror of young and old alike. An aged inhabitant of Penllyne, who died a few years ago, said that in his boy hood the winged serpents were described as very beautiful. They were coiled when in repose, and *"looked as if they were covered with jewels of all sorts. Some of them had crests sparkling with all the colours of the rainbow".* When disturbed, they glided swiftly, *"sparkling all over"* to their hiding place. When angry, they *"flew over people's heads with outspread wings bright, and sometimes with eyes too, like the feathers of a peacock's tail".*

He said it was *"no old story invented to frighten children",* but a real fact. His father and uncle had killed some of them, for they were *"as bad as foxes for poultry".* The old man attributed the extinction of the winged serpents to the fact that they were *"terrors in farmyards and coverts".* An old woman, whose parents in her early child-hood took her to visit Penmark Place, Glamorgan, said she often heard the people talking about the ravages of the winged serpents in that neighbourhood. She described them in the same way as the man of Penllyne.

There was a *"king and queen"* of the winged serpents, she said, in the woods around Bewper. The old people in the early days said that wherever winged serpents were seen "there was sure to be buried money or something of value" near at hand. Her grandfather told her of an encounter with a winged serpent in the woods near Porth-kerry Park, not far from Penmark. He and his brother *"made up their minds to catch one, and watched the whole day for the serpent to rise. Then they shot at it, and the creature fell wounded, only to rise and attack my uncle, beating him around the head with its wings"* She said a fierce fight ensued between the men and the serpent, which was at last killed. She had seen its skin and feathers, but after the grandfather's death they were thrown away. The serpent was as notorious *"as any fox"* in the farmyards and coverts around Penmark.

It is truly frustrating that this priceless skin was discarded - it could well have been the most important zoological specimen of all time! This said, the very fact it was thrown out shows that the populace did not consider the winged serpents anything out of the ordinary. True, they were pretty, but also a pest to farmers and looked on in the same way as a fox or buzzard.

It has been theorised that the serpents were some kind of tropical birds set free from captivity, but no birds that could prey on farm stock remotely resemble these Welsh wonders. Perhaps in some cellar or stock room of a museum, or in the attic of an old Welsh farm, some remains of these remarkable beasts still lie in wait for an incredu-lous scientific community to discover them.

I was due in for an interview with the *Penarth Times* about the mystery cat and the bird like thing (later dubbed the `Penarth Pterodactyl`) so before I went in I measured the *Penarth Times* window. I found the window measured around 2.60m (8.5ft).

During the interview I put forward my theory on what the mystery cat could be, a tanuki or raccoon dog *(Nyctereutes procyonoides)* fit the description perfectly, right down to the colour of the bushy tail. Tanuki, originally from Japan and east Asia, were present in genetically viable populations as far west as France, so it doesn't just eke-out an existence in climates like ours, but it actually survives quite comfortably. They are also kept as pets in some households in the UK, so it was certainly not too much of a leap to postulate that an escaped tanuki could be living in Alexandria Park. There are several areas of the park in the Bridgeman Road end that are almost inaccessible to all but the most determined of people, but would make a perfect habitat for a tanuki.

A tanuki or racoon dog (Nyctereutes procyonoides)

The bird-like creature was a different kettle of fish altogether though, and not so easy to pin down; especially considering the witness had not wanted further contact on the matter, nor left their name with the reporters of the *Penarth Times*, only the road they lived in. However, given the fact that a bird, even one with a wingspan of over two and a half metres, was unlikely to put the fear of God into a person, it was certainly strange.

As one possible explanation to the paper I suggested that the bird could be an example of a zooform phenomenon.

I later questioned a local hawker, Marcell Zalleck, on the subject of large birds of prey that may fit the bill and the only one that really came close to a wingspan of that size was the harpy eagle *(Harpia harpyja)*. Harpy eagles are natives of South America, black and white, and have a wingspan of two metres. They are the second largest eagle

*The graveyard in Penarth, complete with
`winged things` of its own..*

Above: the offices of the Penarth Times
Below: Purcell Road - the haunt of a pterosaur?

in the world ,after the endangered Philippine monkey-eating eagle *(Pithecophaga jef-feryi)* which has a 2.2 metre wingspan. Harpy eagles are on the I.U.C.N. red list of threatened species themselves, so they are not easy or cheap to get hold of in the UK. They are also very temperamental, vicious and difficult to train. The Welsh hawking centre in Barry, near Penarth houses an eagle owl ,but it has never escaped control to run amock across Glamorgan.

In favour of the zooform explanation: if tulpas can be created by high amounts of stress, Purcell Road - where the bird-like creature was seen - would certainly be some-where that a lot of stress would accumulate; slap-bang in the middle of a council estate, and across the road from Penarth's large graveyard. It's not hard to imagine a winged tulpa coalescing into existence among the gothic-looking angel monuments of the cemetery, as generations of relatives wept and buried their dead. There are a lot anec-dotal tales of graveyards and cemeteries being 'haunted' by cowled monks, shrouded figures, and strange creatures; indeed one of the most famous sightings often explained as zooform, the `Owlman of Mawnan` was witnessed in the skies above the church's graveyard and tower.

Harpy Eagle (Harpia harpyja).

The Penarth Mystery Cat

My interview was due to appear in the issue after next of the *Penarth Times,* and in the intervening week two interesting letters appeared; one from the CFZ.'s own Chris Moiser, postulating the theory that the mystery cat could have been a raccoon (*Procyon spp.*) and another from a local lady called Haze Marsh who had seen a large cat matching the description of the Penarth mystery cat, even down to the bands on its tail. The week of publication of the interview in the *Penarth Times* I decided to get photographs of the Purcell Road area, gardens there and the cemetery, to illustrate a planned article for *Animals & Men* about the bird-like creature.

It was here after photographing gardens in Purcell Road, that I came face to face with the Penarth mystery cat himself! Needless to say, to bump into one mystery animal when on the look out for an entirely different one, was perhaps one of the strangest coincidences and examples of pure dumb luck I've ever been party to. The mystery cat, it turns out, was not a raccoon or a tanuki but an abnormally large silver and black tabby tom cat with a very bushy tail, as suggested by Haze Marsh. As I had my camera with me I took several photographs of the 'beast', who was curled up because of the cold from a recent snowfall, and who obligingly posed for in return for a tickle under the chin.

Sadly, at the current time the identity of the bird-like creature remains a mystery, with the two possible explanations being a zooform phenomenon, or a very large raptor, but not nearly enough information was given in the sighting to be able to conclusively say it was one or the other. But if more information ever comes to light, or other sightings are made, then that might become possible. However it looks like `case closed` on the Penarth mystery cat, who may not have proved that mysterious after all, but was certainly an unusual example of a domestic cat.

IDENTIFIED FLYING OBJECT

by
Noela Marshall Mackenzie

A man at the same time both black and white exists, and is easily found. The tincture is not that of greasepaint. He is no actor and is not to be found on the media. He actually is to be found at no greater distance than a portrait from life by a talented artist dating from the 1940s.

His name transliterates as *ia-in-an. Na*, which is the tetragrammaton. Robert Graves brought him into his book *The White Goddess* as Gwion, 'the fair one with the steady hand'. Information on his life and exploits can be found on the Internet. He is Sergeant John Hannah VC, an air gunner with the R.A.F. from 83 Squadron who was sent to Antwerp to destroy invasion barges.

While over Belgium the bomber caught fire, and Sergeant Hannah fought the flames on his own and brought the aircraft back to base. While he was engaged on this task, his face was burnt black, and when he came to earth he was both black and white. Normally he was fair, being nick-named 'Snowy' by his colleagues. For this deed he was awarded the Victoria Cross, and continued to serve as a gunnery instructor. The war artist Eric Kennington had a series of portraits of R.A.F. personalities, and he made a portrait of John Hannah.

John Hannah was not a good subject, and having his picture created was most likely a greater ordeal than battling with fire.

He was invalided out of the service, and died in 1947 of T.B. The Kennington portrait is in the Imperial War Museum, but not on public display.

L-NUMBERED CATFISH

by David Marshall

with photographs by Dr Iggy Tavares

I the late 1980s a small variety of loricarins new to the U.K. aquarium hobby began appearing in aquatic retail outlets. All of these fish were given exotic-sounding common names so a small white fish with black stripes was sold as the emperor or zebra peckoltia, a fish with wavy black and yellow markings the scribbled plec, and one with a dark black body and white spots was sold as the vampire plec.

From the scant information that could be obtained, mainly through friendly retailers, UK aquarists were led to believe that all of these fish had originated from the Rio Xingu area of Brazil and were vegetarian by nature. It would take sometime for this information to be corrected, and make aquarists realise that these particular loricarins' natural range extended beyond the Xingu area and that their dietary requirements were actually very varied.

As more of these loricarins began to appear, the sales tickets on their aquaria (first seen in Yorkshire through L 018 - *Baryancistrus niveatus* 'golden nugget') began to show a sequence that began with the letter L followed by a series of numbers. Shortly afterwards the letters LDA began the sequence with some of these fish. The shape and character of the sequenced species was also starting to change as no longer did they all have the look of a miniature plecostomus, but some resembled whiptail catfish, others large otocinclus, and then came the truly bizarre sight of little loricarins with grey and black marbled patterns - best described as elongated wine gums with fins. What on earth was going on?

Hypancistrus inspector

Bristlenose pleco sp

Confusion in Germany

When all was revealed, it became clear that these loricarins had been available on the European mainland for some time before we had seen them in the U.K. They were appearing not only from the Rio Xingu, but from many other areas of South America. As each new fish had been discovered, the well known German magazine *DATZ* had featured their portrait.

Worried about varying common names that had been given to these fish in the trade, the DATZ Editorial staff had come up with a system in which each new fish would be given an L-Sequenced number that would allow it to be universally recognised, until the scientific community could get through the tortuous procedure of giving them a proper scientific name. Going back in their records revealed that the fish U.K. hobbyists would come to know as the `white spot pleco` had been the first to be pictured, so this fish began the L Sequence as L 001 (L1). With some of the new species, it was clear to see that these were only colour, or regional, variants of loricarins already sequenced. So this was indicated, by adding a lower-case letter at the end of their number. Thus L O90d is the fourth known variant of 'panaque species Peru'.

The Editorial staff of the rival German magazine Das Aquarium had also received photographs of other new loricarins coming out of South America, along with different-looking photographs of some of the L numbered fish, so not to be outdone they created the LDA numbering system for their photographs. Starting this sequence was the gold peckoltia which thus became LDA 01 (LDA1).

So from the start, we had confusion with different L and LDA numbers applying to the same fish. This became even more compounded when regional and colour variations of sequenced loricarins would slip through the system, thus giving them a totally different L number than that already assigned for their kind.

This would become partly to blame for further errors occurring, as other aquatic publications began captioning loricarin photographs with incorrect L and LDA numbers. These problems aside, the two sequencing systems remain the recognised - and best - way to keep a record of all the subject species, until the arduous task of scientifically naming them all is finally completed.

How many L and LDA numbered fish do we know?

It is hard to keep track of LDA-numbered fish through publications available in Britain, but by September 2003 a total of 76 loricarins had been sequenced. Thankfully, information on the L numbered loricarins is easier to come by, and of September 2004 the sequence had reached number 387. According to the Y.A.A.S. Fish Showing Size Guides, a total of 52 L numbered fish have now received valid scientific names, and these are split into a large number of genus classifications. With the LDA sequence, this process appears to be going much more slowly with fewer than 20 of these species having gone through - as yet - the scientific nomenclature process.

The well-known Canadian aquarist and fish collector Oliver Lucanus, while speaking to members of the Catfish Study Group UK, believes that there are many more such fish awaiting discovery, but that this task may become a race against time, as the increasing demands caused by intensive agriculture and search for mineral wealth, threaten the future of many habitats.

Pleco shoal

Golden nugget pleco

On the other side of the coin, a number of communities along the Rio Xingu now make a living collecting just one or two of the most popular L and LDA numbered fish, and such is the conservation concern over this practice, that we have already seen a collecting ban put upon the beautiful zebra peckoltia.

Golden rules

There are four golden rules which apply to keeping L and LDA sequenced fish:

1. Always find out as much information as you can about any of the loricarins that you wish to purchase. Not only do their feeding habits vary greatly, so too do their natural habitats, so we find fish originating from river rapids, lowland rivers, uncharted depths, brackish areas, sandbanks, areas rich in plant growth, and fast flowing rivers where fallen trees provide the cover, all lumped together under the sequencing systems.

2. Never purchase specimens with thin-looking bodies and sunken stomachs, as once many of these species have ceased eating, they may never have the will to do so again.

3. Although the majority of these fish will take standard catfish food tablets, sticks, and flakes, in aquaria, many are specialised feeders. The best guide we have to establishing the supplementary foods needed in order to maintain their health, is by looking at their teeth (which is not always easily achieved).

4. Basically, those with a single row of thin teeth on each side of their mouths, prefer a vegetarian diet. Those with two rows of impressive-looking teeth, bore into wood, so thus need to be provided with mopani or bogwood. Those which appear to have two sets of woodlice imprinted onto their mouthpad, are omnivorous by nature, whilst those with mouths that show a pattern resembling an Olympic Torch, need to be fed a carnivorous diet - including mussels and shrimps.

As we shall see, it is important to do daily checks, if possible, as to the health of all L and LDA sequenced fish.

A Warning

We cannot leave the L and LDA numbers without warning of the main drawback in keeping these particular fish. A number of these loricarins are prone to dying very suddenly and without giving any indications of ill health. When we realise that this has happened, we should remove the body straight away, and make a water change to help clear any pollutant this death may have caused to the aquarium water. Unfortunately many die unnoticed, and this is when the real problem begins, as the flesh of a number of these species can begin to decay very quickly.

As this decay sets in, a 'bacterial soup' is soon formed. This 'soup' badly affects the breathing of fellow tank companions, and when this condition takes hold, it can have devastating effects upon the whole community. Over the years, I have heard a number of accounts of whole aquariums - be they stocked at low or high densities - wiped out through this condition, and - sadly - large water changes and commercial aquarium disease treatments proved no antidote.

Without a properly-conducted scientific investigation, we do not know if this bacteria, or whatever it actually may prove to be, is dormant in the body of a number of these species, or if some of the aquarium foods they are fed - such as mussels - actually cause its fermentation.

Just to reassure our readers, many L and LDA sequenced fish are long-lived, and I had the company of a coffee and cream Plec for close-on eleven years.

In Part Two of this article, we will take an in-depth look at a number of the most popular L and LDA numbered loricarins.

n Part One of this article we looked at how the L and LDA numbering system for loricarins came about. This time we will focus on the most popular L and LDA numbered fish, looking at their care and breeding.

Glyptoperichthys joselimaianus

Ancistrus

Picking out one bristlenose from the sequenced fish was difficult, so I opted for LDA 08 (gold marbled bristlenose or Ancistrus 'species Mato Grosso'), originating from Brazil, which has the scientific name of Ancistrus claro. Growing to 6cm, these fish have a beautiful orange-brown body colour. Males are told apart from females through larger - and thicker - head spines.

LDA 08 prefers a hard water environment with a temperature of 26C. I kept a trio - one male and two females - in the company of swordtails without any problems. Like all Ancistrus they are very quarrelsome, and stake-out territories which are held until feeding time, when the urge to devour flaked foods, algae wafers, catfish tablets, and any brineshrimp, missed by the swordtails brought about a truce in proceedings.

Although these fish are very hardy, and resistant to many aquatic diseases, they are prone to one particular malady - vibration syndrome. When a severe thunderstorm hit the Ryedale area, the thunder caused the shelf, on which the aquarium housing my trio was kept, to vibrate. This caused such panic that I had the heartache of seeing the fish roll over and die in front of my eyes.

To write fully about the breeding procedure and fry care of LDA 08 would need an article to itself: A compatible pair seeks out a cave-like structure. Once the female has spawned, she - rather sensibly - vacates the cave, leaving her mate to guard the orange coloured eggs. An overactive male can do great damage to a female spent of ova, so this is something we must taken account of. About a week after the eggs hatch, the fry will be seen scurrying over the substrate and glass etc. feeding upon algae.

We help their growth through feeding crushed algae tablets, boiled nettles, and by trying to get them to eat live brineshrimp. Unfortunately for those of us who prize the natural forms of fish, some of the sequenced Ancistrus are already showing the signs of commercial breeding programmes, and currently available are butterfly forms of the xanthic-looking L 144 (which is much easier to breed than LDA 08) with such large fins that they find manoeuvring around their aquaria very difficult.

Chaetostoma

This genus is home to the bulldog plecs. So many bulldogs, of various sizes and colour patterns, are arriving in the U.K. at the current time that they appear to have by-passed the L and LDA numbering sequences. Of those which are sequenced, L 188 (white spotted bulldog), and LDA 11 (marbled Mato Grosso bulldog), have lovely body patterns upon importation but, as with the majority of their genus, these patterns often fade to an overall muddy green colour as the fish begin to age.

There are two very important factors to be considered when keeping these fish. Firstly, always quarantine any potential new tankmates, as all bulldogs are prone to whitespot disease, and can become so badly infected that you cannot see the flesh for spots. Secondly, always keep bulldogs as single specimens, as members of a group will often, unseen; wear each other down to a situation where only one survives.

Much debate has ensued about how these fish - some of which are found in brackish waters - should be kept in aquaria. From my own experiences they prefer an aquarium no larger than 60x30x30cm, pH7, airflow just above normal, and plenty of regular water changes.

All standard aquarium foods are eaten with great gusto. I have found that small barbs and platies make good companions.

I have only come across one spawning report for these fish. The aquarist concerned kept two bulldogs in separate aquaria side by side. One day it was discovered that one of these fish had jumped into the others tank, and, when found, was seriously battered. Concerned about the health of the resident bulldog, a search ensued, with the aquarist having the surprise of finding the fish - assumed to be a male - guarding eggs in a crevice between two rocks.

Cochilodon

L 77 (coffee and cream plec.), L 137 (rusty plec.), L 138 (black-spotted 'Bruno' plec.) and an un-sequenced Cochilodon with blue eyes, are all recognised under the tag of Panaque species 'Bruno'. Of all the L numbers, L 77 is probably the one most prone to whitespot upon importation, but - thankfully - this is easily treated by using the old method of raising the water temperature, and scrupulously siphoning the gravel.

When a scientific classification comes along, it may well be that all the four 'Bruno' species are given the same scientific name, because they are so alike in features, that only slight differences in body and eye colour tell one from the other. These fish originate from Brazil, where they reproduce inside the trunks of decaying trees, and can reach a size of 30cm. Although their sucker-like mouths are adapted to chewing at wood, they take standard aquarium foods with great enthusiasm. Keep at a pH of 7 and a temperature of 26C. They will accept various tankmates from Corydoras through to large Synodontis.

Hypancistrus

This genus was erected in order to accommodate the beautiful ice-blue and white striped fish Hypancistrus zebra (emperor or zebra Peckoltia). Although there are several other similar shaped and coloured L numbers, the true zebra is L 046. Also worth looking out for is the queen arabesque, L 260, which is widely tipped to make an official appearance in the Hypancistrus genus in the near future.

Panaque

Of the several colour forms of *Panaque nigrolineatus* (royal or pin-striped Plec/Panaque) which carry an L number, and derive from Southern Columbia, my favourite is L 191 which, when young, has a brilliant black-coloured background to its body. Although capable of growing to 25cm, those seen in aquaria rarely reach this size.
An aquarium of at least 90x30x30cm is needed to house a nigrolineatus. They make good tank companions, but dislike the company of their own kind. Although soft, slightly acid to neutral water is recommended, they will tolerate some deviation from this. Keep at a temperature of 25C. Bogwood or mopani wood must always be included in the set-up, as they take various enzymes from this wood which aid digestion. These fish need much vegetable matter to be included in their diet, so we turn to algae wafers, and vegetable-based flake foods

Panaque have the strangest life expectancy of any loricarin that I know. As their teeth are worn down, through munching at wood, several new sets are regenerated. Once the last set is used the fish are no longer able to feed so - sadly - starve to death. Their end, therefore, comes not through age but through how much wood they have consumed.

Leopard Frog Pecktolla sp

Pseudancistrus

To show how both the L and LDA systems can have numbers relating to the same fish, we will talk about the beautiful *Pseudancistrus leopardus* (leopard Acanthicus) identified by the numbers LDA 07 and L 114. They come from fast running water courses, and are natives of Brazil. Depending upon their mood, the plec-like body shows either a yellow or orange background, with a beautiful foreground of black spotting and bars. The tail is a bright orange-red.

Although I have not tried this, I am reliably informed that it is possible to keep several of these fish in one aquarium. This aquarium would need to be of a fair size, as these fish can reach a total body length of 35cm. Filtration needs to be of a high quality, as the catfish will start to fade away if their aquarium is not in pristine condition. Keep at a temperature of 27C. Although these fish are primarily vegetarian, taking lettuce and pieces of raw potato. They will take commercial catfish pellets, and large-sized flake foods.

The sexes are distinguishable, as the edges of the pectoral fins are thicker in males, who also tend to be the more aggressive. We have few pointers as to how these fish actually reproduce, but it is believed that they may follow the Cochilodon way of making nesting sites in the wood of decaying trees.

Finally we must mention the loricarin species sold as L 128 (blue Pleco.) and L 200 (green Plec.). These fish, which originate from the Rio Orinoco, have become the basis of cottage industries for several villages, with young boys diving to great depths in order to catch the best specimens. There was a time when all of their catches were exported almost exclusively for the Japanese aquarium market, but now these beauties are appearing more often in the U.K. So close in characteristics are these two forms, that in the near future they may both carry the same scientific name.

BOGEYMEN OF
ARGENTINA

by
Neil Arnold

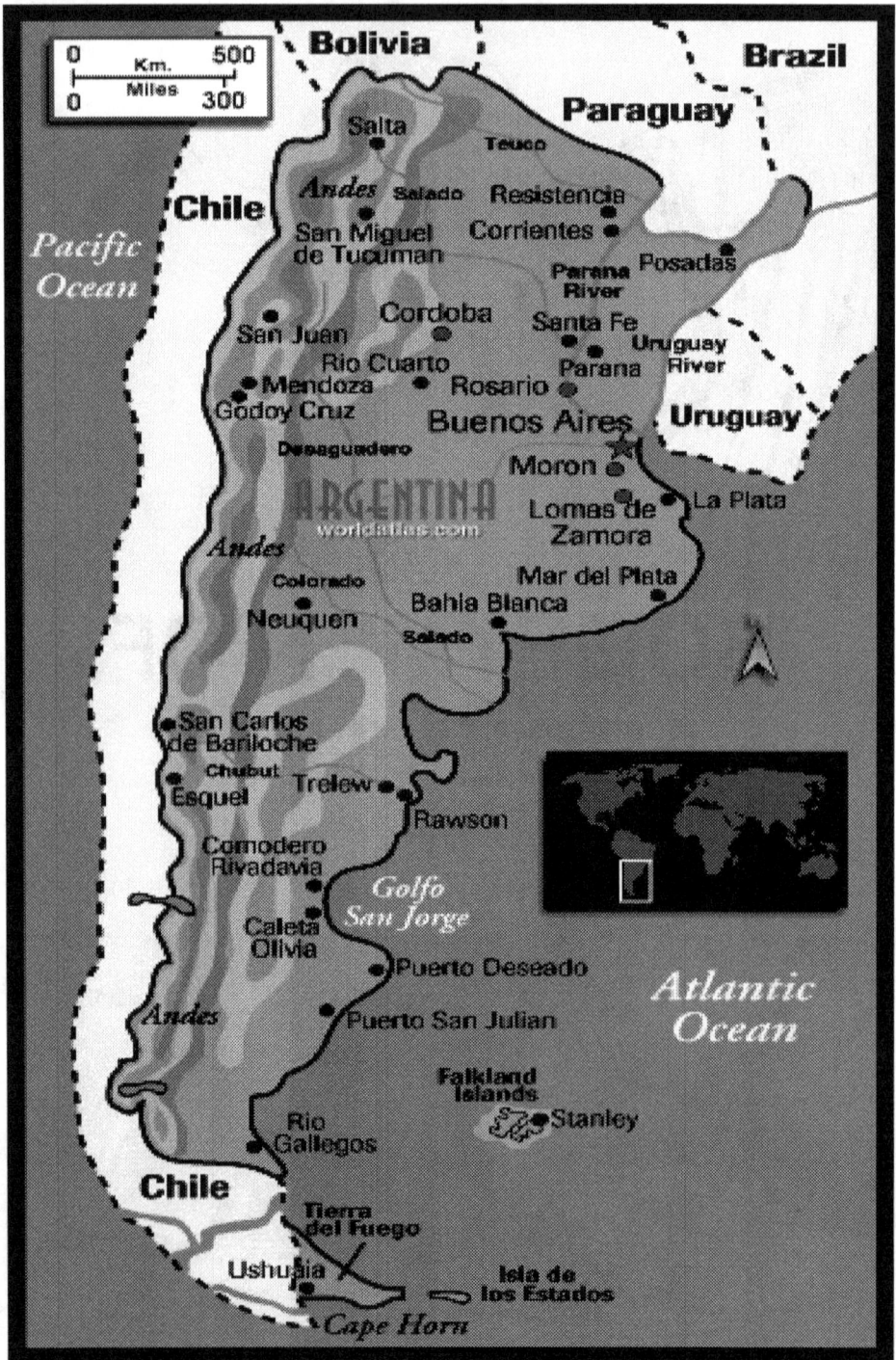

NEWS FLASH!

February 4th 1965 – Torrent, Argentina.

Several witnesses have just observed five luminous objects fizz across the sky. Shortly after the brief sighting, a transparent craft landed in a field, and five creatures, approximately two-metres in height emerged. Witnesses noticed in horror that these critters had a single eye on their forehead and wore flashing helmets.
The imps allegedly have tried to abduct a villager.

NEWS FLASH!

February 21st 1965 – Chalac, Argentina. 9:00 pm.

Over forty Toba Indians as well as police have watched three imp-like men emerge from a strange craft, that for several hours has been zooming over the village. The humanoids, according to some of the witnesses, appear to have luminous glows emanating from a section of their bodies. A man has tried to take a photograph of the strange invaders who seemed horrified by the camera flash. Those close to the scene said that the 'little men' scurried back to their craft which increased in its luminosity and then took off at great speed.

NEWS FLASH!

April, 1965.
Time – unknown.
Location – Monte Grande, Argentina.

Local man Felipe Martinez, 37, reported that he was paralysed during the landing of a silent, large, egg-shaped object from which emerged a small man, about one metre tall, wearing a helmet linked to the craft by three cables. The being spoke in Spanish then left the scene.

High strangeness in Argentina, and South America in general, is common-place, although judging by press reports and media interest, what we have come to know as the 'chupacabras' has only existed in the human subconscious for the last thirty years or so. The facts though are frighteningly consistent, the activity worryingly bizarre, and the truth far stranger than the media fiction.

Reports of Unidentified Flying Objects have existed from the early 1800s. Argentina has always been part of that hot-bed of activity which, although worldwide, seems eerily focused on certain countries within South America, i.e. Brazil, Chile. The mutilation waves are just as frequent, and the dreaded 'goatsucker' that terrorised much of Puerto Rican folklore, has shown its face on many occasions in the local fields and ranches; or so they say. And there is so much *more* thrown into the cauldron for good measure. The folklore is varied, colourful and extremely weird, from werewolves to Bigfoot, from tall, shadowy figures to odd imp-like manifestations. And this article will concentrate on the more obscure mysteries of Argentina, and some of the surrounding countries. What I aim to do is to show you, dear reader, that there is more to the *Chupacabras* and the likes than meets the eye, and that

a)   they will never be proven to exist,

b)   they probably don't exist,

c)   that they have existed in some form for thousands of years.

Confused ?

Once again it's that mention of 'zooform'; the word coined by Jonathan Downes to describe creatures that quite simply *aren't,* or maybe never will be. despite many sightings. Zooform is that word that most crytpozoologists will come to dread, or simply ignore, because they strongly believe that what they are pursuing is flesh and blood , when the folklore clearly suggests it isn't.

Folklore is often scoffed at as rumour, as tribal whisper, as town gossip, and urban legend, but it's these rumour mills that keep mysteries, real or not, very much alive. And it's also these spun yarns that prove that such bogeymen of nightmares are in fact real in some sense, even if just in the minds of frightened kids. That's what zooform is all about, but I do not expect you to understand it, just read it.

For the world of zooform will not uncover any corpses, for it is a vague territory, an obscure place of fleeting glimpses, hysteria, campfire whisper, and closet monsters. It is a place which projects our fears, holds our fears, and *takes* our fears, and seems very real. At times the things that are cast forth from this murky void cross into other areas, areas which I've never always liked to cover - ufology, the paranormal; because these are territories also very much misunderstood in a world where so many believe that discs are flying in from the cosmos, and orbs are the spirits of the deceased. But there are times when such ethereal monsters are connected to UFOs, or indeed the realm of 'ghosts' as we like to call them. Who knows? Maybe all these minor mysteries are part of one big puzzle, or maybe they are very much their own complex being. Either way, there's no point trying to understand, but just stand back and watch in awe, because these things happen…more so when millions of people believe in them and *make* them happen without ever realising it.

The three brief cases mentioned at the beginning were taken from Jacques Vallee's important work, *Passport to Magonia*. They were just three, out of *hundreds* of examples I could have used to start this piece. Here are a few more to set the pace.

- On October 1st, 1965 in Aguas Blancas, Argentina, three young students were attacked, in broad daylight by three green-skinned imps. Santos Vallejos, Antonia Aparti and Adela Sanchez were walking to General San Martin School when the creatures emerged from the undergrowth and ran after the youths. Fortunately for the students they were too fast for the critters and reached their school entrance hysterical.

- Three years later on August 31st, 1968, during the early hours of the morning at Mendoza, three witnesses, two of them Casino employees, were stunned to see five 'dwarves' emerge from a strange craft planted on the ground. The humanoids had oversized heads, seemed to be doing something on the ground and then made off in the object.

Almost hundreds of identical reports over the last two hundred years have emerged from Brazil, Chile, Venezuela, Uruguay and Paraguay, all via the press, Internet sources and literature. Imps have been invading these parts for many years, some from strange flying objects, some from the forests, but in general, from parts unknown, but probably from the darkest reaches of the mind.

In the UK these kind of critters are pretty much forgotten relics of lore. Sprites, bogarts, brownies, spoorns *et al*; all mischievous beings associated with distracting weary travellers on the eerie marsh land, and frightening children to prevent them being naughty. Such `critters` exist the world over. However, in Argentina and most of South America they are considered evil, often associated with the grisly cattle mutilations, allegedly spawned from the Devil himself, or having origins from some distant planet, or, as in the case of the *Chupacabras*, a vampyric entity said to slaughter livestock, cause mass hysteria and also be a symbol of poverty. They are very much real, but they don't exist. Fact.

The *Chupacabras*, or 'goatsucker', has destroyed many hen pens across Argentina, left animal carcasses exsanguinated in remote parts of Chile, and has even been seen emerging from space craft in parts of Brazil. Many have seen it, they'll even tell you about how it levitates, how it has large eyes, and can appear like a flying monkey, or a baboon crossed with a kangaroo, but every time they'll tell you it's the 'vampire'.

These things are all the same but different.

For now, let's leave much of the cattle mutilations and flying saucers out of this complex mystery, because they are very much huge mysteries themselves, so we can concentrate on the matter at hand; this being the strangeness that is the impish figures, vampire dwarves, and dastardly `critters` that have caused havoc in these humid parts, plus one or two other demonic shadows.

On May 27th, 1958 at Boca del Tigre, Remo dell'Armellina was driving his truck en route to Santa Fe, when he saw a figure standing in the road. As he approached, he estimated it stood around three-metres, and appeared very threatening. Thinking something strange was in the air Remo stopped his vehicle, grabbed a metal bar, jumped from his cab and walked towards the apparition. Immediately the pungent odour filled his nostrils and a blinding light threw him to the ground.

The next thing Mr. Armellina remembered was driving again. Only later, once the shock had subsided, did he recall that the figure was covered in scales, had long arms, and wore a tight fitting flight overall.

On the night of September 22nd, 1967 at Caracas, Venezuela, a race track employee coming out of work was grabbed from behind by a creature he could not see in the darkness. The invisible marauder attempted to choke the victim, but a horse neighing in a nearby stable distracted the attacker. Thirty minutes later another employee saw a bizarre dwarf-like creature 'zoom' out of a building, and then attack a horse. This echoed a similar attack two years later in Buenos Aires, when a night watchman claimed that whilst on duty he was assaulted by a three-feet tall, dark-coloured `critter`.

Many peculiar, and at times frightening encounters with dwarf-like entities that have been recorded from several decades ago are extremely brief. They are brief almost to the extent that they are forgotten ditties, almost comical, maybe at times found lurking within the pages of some UFO-related tome, but their importance to this article is not underestimated. They must not be underestimated, because these creatures, entities and critters are very much part of Hispanic culture now, with great fears and dreads among the communities, especially since the *Chupacabras* attacks have come to the fore.

The most feared creature of Argentinean lore is *Ell Diablillo*. It showed its face on 5th July 2000 five km from Valle De Catamarca, near Dande de Varela, when it appeared before two police officers at their substation. Corporal Miguel Angel Aguero was asleep along with Luis Rodolfo Aguero and Walter Ortega, when they were disturbed by a sound coming from the roof of the next bedroom. Ortega decided to go outside and check out the 'stamping noise', but as he entered the room, he was lifted into the air by an unseen force and thrown up against a tree violently.

Angel Aguero summoned a back-up driver to the scene and told all he'd seen a small figure with red, glowing eyes, wearing a green shirt with black trousers! Aguero claimed that the creature said to him, *"...I come for thee on Satan's behalf"*.

Such a report may sound like something from a cartoon strip, but the amazing thing is, so many reports of impish figures often concern creatures dressed in tight fitting shirts, trousers, at times wearing hats, but having either glowing eyes or ugly skin.

On April 17th 2000, at Frias, 162-miles from Cordoba, another police officer encountered an imp, this time by the roadside, which at first he took to be a small child ,and so approached it without concern. As he got closer to the figure it turned to face him and revealed two, red glowing eyes.

Ell Diablillo, when translated from Spanish, reads 'imp from Hell'.

On February 22nd 2004 another 'imp from Hell' began to terrorize parts of Argentina, this time at Saenz Pena (Agencia), and was witnessed by 26-year old Liliana Nieves who was allegedly attacked by the horrid entity.

El Norte Digital reported:-

ARGENTINA – WOMAN NEARLY
BATTERED TO DEATH BY 'IMP'.

Nieves, claimed from a ward at the *4 de Junio* Hospital, that a small, black-coloured man in a black hat had battered her and attempted to drag her away.

It all began on a Tuesday when a shower of stones fell onto the home of the witness in the tranquil area of 31 & 0000 Street in Santa Teresita. The lady claimed she could see a small figure tearing at tomato plants in the yard, and when bricks began to land on the roof, a shadowy figure was observed to flee from the area. At first Liliana thought it was a prank, until the Thursday night when she and her husband were once again awoken by a shower of stones on the roof. Liliana went to investigate, but never returned. She was found a short while later by a local man; she was lying on the ground, her face beaten ,and signs on the ground that her mystery attacker had attempted to drag her away.

Neighbours in the area stated that the attacker was invisible, but many locals believed that the creature was an imp, a creature of folklore mentioned in songs and said to lurk in parts of the river basin. To the Araucan natives, there is a similar 'critter' known as the *Peuken* said to inhabit the dense forests of Chile, whilst in Argentina the *Pombero* is also known as the *Pirague*. Such mystery figures are often perceived as bogeymen, they are common through all world folklore and belief systems, i.e. the *Eloko* of Zaire, the *Popobawa* of Zanzibar et al.

The small, nocturnal creatures of South American lore seem to have two different motives. The imps, being humanoid figures, appear to target people, whereas the small, animal-like critters are often associated with attacks on livestock.

The Cordoba region of Argentina seems to attract, or indeed spawn, varying imp-like terrors. One was sighted, date unknown, by fourteen-year old Juan Pablo near Aguas de los Palos. He described the thing as short, carrying a strange staff, and wearing a red jumpsuit and cap!

The larger figures of Argentinean nightmares are rarely seen, but they exist. The Monte Maiz phantom may be far removed from some grinning imp, but it has the same eerie motive; to terrify the public, to seduce and assault women, motives echoed during the late 1800s by London's own Victorian marauder, Spring-Heeled Jack.

The Monte Maiz phantom has existed in the minds of the locals for many years, but once again, encounters are scarce, but when they *do* occur ,they embed themselves in local history like thorns in the side.

In April 2004, Scott Corrales from the Institute of Hispanic Ufology stated that, *"...the peace of this 9,000 inhabitant locality in the department of Union, was interrupted in the past three weeks by a strange creature who wanders the night to terrify residents. The residents call it 'the phantom', those who have seen it say it is a young person, very tall and slim, with red eyes."*

Descriptions, as always, and certainly in the eye of the hysterical storm, often vary. Many residents to the spectre claimed it was a humanoid figure in bright white clothing, but others believed that what they'd seen was a spook in dark garments and a cape, sometimes wearing a black hat. Some consider it an urban legend terror that whispers words of death into the ears of those who encounter it, whilst women fear it is a nocturnal rapist, keen to drag them from their slumber into the dark of night.

Monte Maiz though, like many other areas of Argentina, is known for its UFO lore. During the 1960s hundreds of flying saucers and strange disc-like craft were spotted in the skies over the town, but there's certainly nothing to suggest that the caped intruder popped out of a mother ship! If we begin to connect such phenomena, then we have an endless stream of possibilities that only confuse an already complex mystery, and the likelihood is, there is probably no connection anyway. But what you will find in much of the paranormal world, is that extraterrestrials and UFOs are often blamed for many bouts of high strangeness in certain areas. Crop circles, cattle mutilations, strange creature sightings, Men In Black figures, black helicopters, etc., have often been connected to clandestine alien activity, but why these relations come about we'll never know.

Idealised image of Lobizon from an Argentine website

Sometimes things are just plain weird, but to suggest they were put there by cosmic visitors is even more absurd.

El Petizo is very much a comic book character—also from Argentina's weird lore. The town of El Duraznito has been targeted on several occasions in the last few years by what is known as 'the shadow'. The elusive figure prowls the northern Salta province, preying on unsuspecting victims, even though it has often been perceived as nothing more than a tall tale told around too many campfires to scared youths on dark and stormy nights.

Researchers believe the creature comes at certain times. In 2002 the dark figure of menace attacked a boy on his bicycle, in an area nine miles south-east of Rosario de la Frontera. The attack took place in daylight, however, when the boy was en route to visit friends for an adventurous day of hunting. With his knife and shotgun packed, he made his way to his pals ,but was suddenly thrown from his bike. Lying on the ground, he turned to see a shadow on the road approach slowly. The boy bravely reached for his gun, and fired at the presence; but to no effect. The sinister apparition pushed the boy to the ground time and time again, and then clasped him by the hair, and attempted to drag him off into the woods. Thankfully for the victim ,his screams were heard by a local man who came to his rescue, just in time to see a weird shadowy, human-shaped blackness pulling the boy along. As the man went to grab the boy the shadow vanished.

This specific case was blamed on local occult practices alleged to have taken place in the cave of the 'black ones' occupied one-thousand years ago by the Candelaria cult!

Argentina's other most prolific entities of aggravation and horror are without doubt the 'hairy ones' such as the *Dientudo* and the *Lobizon*, werewolf-type bogeymen said to inhabit the thick creeks and even dark corners of the towns at night. The *Dientudo* may be a relative of the Sasquatch, a tall, robust hairy hominid said to have characteristics of a bear but appear as a fearsome humanoid.

On 28th July, 1968 in Prana, Argentina, a student was allegedly attacked by a large, three-eyed monster covered in white, shaggy hair which grabbed him as he drove his motorbike through the district. The witness said the attacker was large and muscular,

Dientudo in popular culture

something akin to a werewolf, or Bigfoot-type hominid.

The *Lobizon* also fits into some of the *Chupacabras* attacks regarding some descriptions, most of which came about after the dawn of the new millennium in areas such as Entre Rios, where animals were found either stripped of flesh or showing unusual puncture marks, the carcass bereft of blood. At the same time local farmers reported seeing eerie lights in the sky over the neighbourhood but there was no real sign of the mystery marauder, but even when there were brief glimpses of the attacker, reports varied. Some terrified locals claimed that what they'd seen was, *"...like a monkey but having a nauseating odour",* another witness claimed it was, *"...a shaggy dog-like animal with a long tail",* and one woman, who said that the beast had killed her pet cats believed it was, *"...an inhuman shape that floated across the yard".* Such activity has also been connected to cattle mutilations, although across the U.S.A. such nocturnal surgery is blamed on government/ military operation or extraterrestrial behaviour.

These manias and scares are very common, often blamed on local poverty, political upheaval, covert operations by the government and impending misfortune such as earthquakes or drought.

The *Lobizon* has often been spotted raiding livestock pens in Concepcion del Bermejo. In the last few years an animal, described as 'dog-like', has been shot at, beaten, and had its hair analysed but to no avail. Such a creature also bears some kind of resemblance to others creatures of South American folklore. The 'winged dogs' of Chile have also been connected or at least confused with the *Chupacabras*, or indeed the *Lobizon* proving that many mysteries may well explain a single mystery that doesn't exist, or maybe none of these 'ghosts' exist at all but the locals are looking for some kind of bogeyman to blame for their current climate.

On July 11th 2004 at Paso Hondo, Chile, peasant Juan Acuna Periera claimed he was attacked by a weird creature after investigating a commotion in his yard, and several deaths of chickens. He told local police that a 'winged dog' was to blame, a creature also sighted a year previously, and blamed for hundreds of attacks on hens in the area. The creature was said to be grey in colour, levitate, and look like a cross between a canid and a baboon. Some of the original *Chupacabras* attacks in Puerto Rico mentioned a similar creature, although it must be said that the typical image of the *Chupacabras* seems to be something projected into the minds of witnesses as reports previous to 1994 very rarely mention the creature that everyone has come to know as the *Chupacabras.* When the 'goatsucker' was originally sighted around Moca in the 1970s, many witnesses reported cat-like animals, or strange dog-like creatures. However, varying other descriptions over the years have been used, and collated, to form an identity that may well be false of the real culprit, but which now has become the true face of the mystery predator. This kind of thing has also happened with other zooform creature.

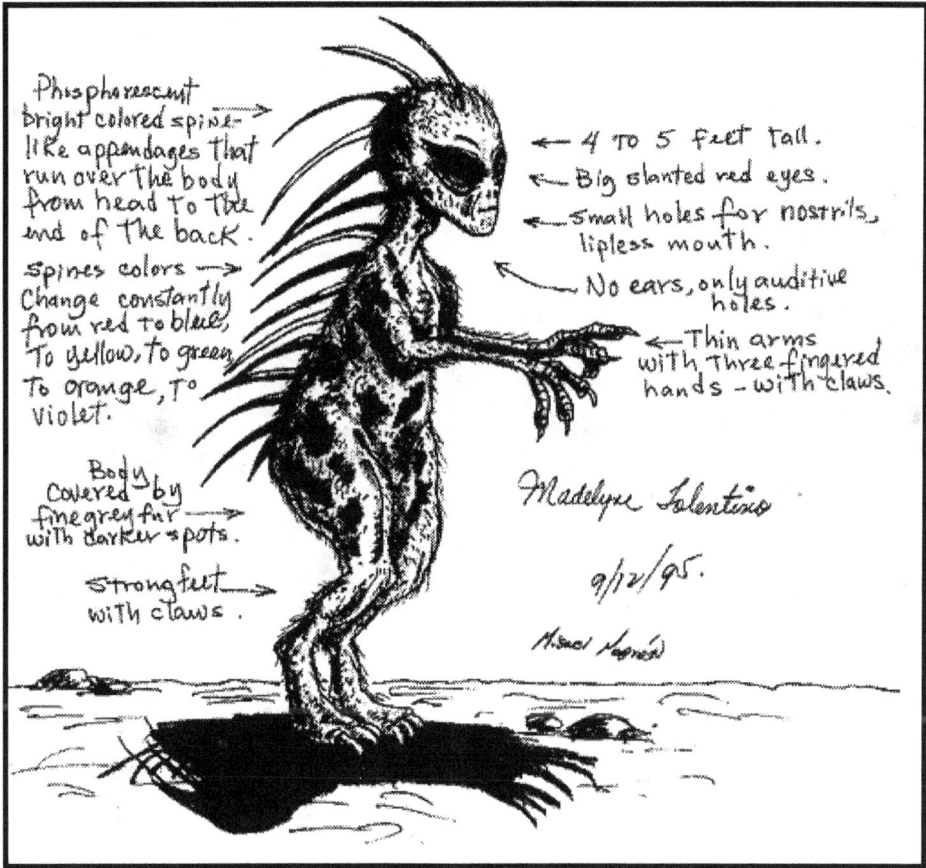

Phosphorescent bright colored spine-like appendages that run over the body from head to the end of the back.

Spines colors → Change constantly from red to blue, To yellow, to green To orange, To violet.

Body covered by fine grey fur with darker spots.

Strong feet with claws.

← 4 to 5 feet tall.
← Big slanted red eyes.
← Small holes for nostrils, lipless mouth.
← No ears, only auditive holes.
← Thin arms with three-fingered hands - with claws.

Madelyne Tolentino

9/12/95.

The best-known image of the Chupacabras

tures over the years such as the West Virginia Mothman who began life as a humanoid with leathery wings but was transformed by history into a black, headless monster with two red, glowing orbs for alleged eyes. The New Jersey Devil began life as a demonic campfire legend but sprouted leathery wings and took on the form of a kangaroo-cum-horse, and eventually ended up as a silvery cat-like animal that preyed on livestock.

What I am trying to say here, is that communities mould these apparitions, they fit them into their own suggestive theories and personal nightmares, their fears and their dreads, and that's how these things take shape. This has been proven with some of the more recent *Chupacabras* evidence, mysterious carcasses that have turned up in remote areas which witnesses claim are of the 'goatsucker'.

On October 14th, 2004, farmer Devin McAnally of Pollok, Texas, shot a strange animal that was munching his mulberries. He shot the animal because he believed it was the

The Texas beast - not a Chupacabra

same animal that had attacked his chickens over a short period.

Allegedly there have been several livestock attacks in parts of Texas as well as Florida attributed to the goatsucker predator, but once again, descriptions vary and the shot creature was ultimate proof of the confusion.

There had been several reports of a creature with a long tail entering hen houses in the death of night, and picking off victims with silent ease. In Sweetwater, Florida, some locals blamed a creature that was covered in shaggy hair, walked upright and has become known as the Skunk Ape, a creature said to prowl the river bottoms, but which appears more like a smaller, more thickly coated Bigfoot, or upright walking Orang-Utan, than any kind of winged vampire.

The animal shot by the Texan farmer (see picture) is clearly not a Chupacabras, or at least certainly does not fit into the known descriptions of the Puerto Rico/ South American goatsucker which appears to have a spiny back, large eyes, claws and smell pungent. The Texan critter, possibly a hybrid creature, coyote etc, is far from a fearsome blood drinker, and yet the locals swore that this was indeed the Chupacabra.

Similar beasts exist all through South America, but when they are sighted nowadays, they are called Chupacabras.

The Viluco Monster is said to inhabit the mines in the Cerro Cullipuemo locality of Chile. Lore of the these shadowy apparitions dates back to the 18th century, elusive, kangaroo-like entities with wolf snouts said to reside in the dark confines of the abandoned mines which have been empty since the 1950s. These little creatures have been blamed for the deaths of more than two thousand rabbits in the Isla de Maipo sector and legend is rife pertaining to these eerie predators. Strangely enough, folklore of Chile also speaks of the Alicanto, a trickster entity said to guard old mines and feast on silver and gold instead of the flesh of local fauna.

The *Buin Monster* is another Chilean entity said to have the muzzle of a wolf, a humped back and reflective red eyes. This creature was reported in January 2004:

"PROSECUTOR, SPOUSE SEE MONSTER!" – *La Cuarta* reported.

Hector Cossio wrote, *"...a state prosecutor and his wife have returned to Santiago after having spent a week in La Serena, where they ran into a hairy monster that made their blood run cold as they took an alternate route to the main highway. They say it wasn't a creature made by the Lord. The beast was nearly identical to the one seen in early January by driver Juan Berrios in Buin.*

Roberto Ayar Rojas, Maribel Arnaiz Cazaux and their young daughter Daphane left La Serena around midnight on Wednesday toward Highway 5 North. 'We wanted to take our time and take scenic routes through inland towns, partly to avoid the recurrent tolls and traffic and partly discover non-tourist destinations, even if it was already night time."

At 6:00 am, as they took an alternate route, on a road flanked by vegetation, they saw the 'thing'.

"It wasn't a dog, rabbit or any other known animal. It was halfway in the middle of the road, standing on two feet. It was completely covered in hair and had red eyes", said Roberto a day after the incident.

They described the beast as like a kangaroo in body form, with two small hands and deep-set eyes. The couple were struck by its gaze and the two long fangs protruding from its jaw.

It was also claimed that around the same time, hairs, possibly from a similar creature were being examined by the GEO Group, but witness Roberto Ayar Rojas had only one word to describe the creature, *"...supernatural."* The hairs examined were said to be over ten centimetres long, black on the ends and white in the centre. They were taken from the vehicle of Juan Berrios who allegedly ran into the beast on January 5th 2004, whilst driving in the same eerie alley that the family had seen the creature.

Berrios claimed that the creature brushed his microbus as it leapt across the road, leaving strands of hair on the windshield. He described the animal as over a metre-and-a-half in height, but with black eyes. He watched the thing for more than twenty seconds, noticing how its muzzle was longer than that of a wolf and it also had a hump on its back."

The Buin creature may well be the same animal shot in Texas during October 2004, some kind creature as yet undiscovered by science, a bizarre mutation, but the goat-sucker ?

The shadowy legends of Peru and Chile also speak of the *Chonchon*, which in fact is actually a flying, vampyric head said to feast on humans. How are such legends born? Could a critter with glowing eyes, levitating in the darkness, so many years ago, some-how be mistaken for a flying head ? Maybe.

The *Chivato* is another ancient Chilean legend. During the 1880s, a magician allegedly on trial, told judges how he had visited the royal cave of Quicavi, via a secret passage-way, which belonged to the house of the present (at the time) sect leader Jose Merriman. In the dank cave the magician said he encountered hideous guardian ani-mals, the *Chivato,* which were covered in hair and crawled low to the ground.

Brazil is also known for its goatsucker-like manifestations that have terrorised locals and their dreams for many years.

The *Bicho Papacirco* is a bogeyman figure that preys on children. It is a vague creature with no real face, but each person who fears the monster has their own image of its grim façade. The reptilian *O Bicho* is a spindly forest dweller said to attack livestock in local villages. The elusive predator comes under the blanket of night, raids pens and remains relatively unseen. Those who have caught a glimpse of the beast believe it to be dark green-brown in colour. The *Jaracaca* is a more ominous relative, said to prey on mothers and their new born babies, a common theme among oriental vampires which, with long straw-like tongues, can pierce walls, or filter through tight cracks in order to seek out the blood of their prey. The *Kuru Pira* is a bigger beast, a red-eyed manimal with a shrill cry, backward turning feet and similar to a werewolf. Prey of the *Kuru Pira* will have its skull punctured by large teeth, all liquid and crimson drained.

In Mexico the beast of Mochis, blamed for hundreds of attacks on chickens and ducks in the Sinaloa vicinity, is said to resemble a small dragon, have protruding fangs and be greyish in colour. Even as far away as Costa Rica there are vampires. In March 2004 in the Ajuela Province, a man and his son were awoken one night by a commotion out-side, centred upon the henhouse. Upon investigating the disturbance, their torch beam picked out a black creature, said to have a long tail, that resembled a dog but stood on two legs. The creature was fleet of foot and escaped its pursuers, but left twenty slain chickens, each being exsanguinated and showing two puncture marks on their backs. Similar attacks have occurred for centuries in Zambia, Peru, parts of Australia, Spain and Germany. How many goatsuckers are out there?

Some would argue that I'm being sceptical towards the existence of the *Chupacabras* and friends, considering the hundreds of reports from local people across much of South America, parts of the U.S., Costa Rica, Puerto Rico etc, and also there is the grisly evidence, in abundance of livestock kills, the bizarre attacks on goats, dogs, and humans some have claimed. However, my point is, these things have existed for such a long time, centuries, even farther back, and many of these creatures have been forgotten, but they often are conjured up again, because weird events *do* occur that provoke us to dig deep back into our folklore in the hope of finding clues as to why these things may happen. Sure, there are flesh and blood creatures out there to blame in some instances, but there is also the hysterical public, fuelled by their common dreads, and emotional fears. For these fears can indeed creep from the forests, crawl into the backyards, and indeed siphon blood from your livestock in the same way a poltergeist can throw a chair across the room. However, I do not believe that we are dealing with real, flesh and blood monsters, and the eye-witness reports prove that to me.

These imps and nocturnal predators have no message from the stars, or from Satan either. They lurk on some remote void, waiting to be projected by a grouping of minds and nightmares. The vampires out there, in reality are nothing more than elusive cats such as the Florida 'panther' or the jaguarundi, but the next day they may well be a winged entity, and the screams in the woods become the wails of that particular common dread in the same way those slaughtered hens become the next batch of victims from a bloodthirsty invader. Those cattle that lay in the grassy fields of the Argentinean plains, stiff as a board; they were once the prey of the great Gaucho bird phantom, the same feathery, beaked beast said to have terrorised local ranchers. Yet, over time the cattle mutilators have transformed. Authorities claim natural predators, but local ranchers and investigators believe it's a cover-up, a military exercise of a sinister nature. And there are those who believe it's a vampire. It's none of the above. For we know that surely, there's no way on Earth, than a green imp, dressed in a red hat and jumpsuit, can appear out of nowhere and deliver an evil message.

On the night of September 11th 1965 in the town of Guarulhos, Brazil, a male witness named Antonio Pau Ferro watched in disbelief as two unidentified flying objects landed in a field. From these craft two beings emerged, both approximately 70-cm in height. They were humanoid figures but covered in 'ugly' and dark skin. Strangely, these `critters` were not interested in mutilating cattle or drinking the blood from chickens, but instead took interest in a patch of tomato plants, before slipping back into their ships and whizzing off to some unknown place. Weirdness happens all the time.

On the night of December 10th, 1954, at Chico in Venezuela, two young men watched in amazement as a large red orb descended from the black sky, and landed near the Trans-Andean Highway. The two men, fascinated by the object which was shaped like two bowls stuck together, approached cautiously. The next thing they knew they were apprehended by four critters covered in hair. The dwarves were very strong ,as they threw the men to the ground before stepping back into their mystery craft and speeding off.

These peculiar dwarves, goblins, critters, sprites, tricksters, invaders, whatever you want to call them are common. However, they are common because there are millions of minds to create them. Just like viruses that aren't, old wives tales, urban legends, panics, ghost stories, lies and scare stories in general. These things filter through and we do not understand them, but we believe them. Even when we fear them so much.

THE SEAL SERPENT

The Case For the Surreal Seal

by Robert Cornes

FOREWORD

Surreal ;   Strange, bizarre.

(Oxford English Dictionary)

I can honestly say that this work has been a labour of love.

The inclination has always been there, but the motivation for a number of reasons, has been sadly lacking. It has therefore been a supreme challenge to finally put ink to paper and reflect on such an unusual and intriguing subject.

After having read Peter Costello's, *In Search Of Lake Monsters*, many years ago as a teenager, I became unusually preoccupied with the rather surreal notion of 'The Long Necked Seal', which was presented as a possible explanation for at least one type of 'Sea Serpent' or 'Lake Monster' found the world over.

At one time quite popular, this theory today is very much out of vogue, having been replaced by the resurgence of that cryptozoological 'darling' the plesiosaur, and to a lesser extent the zeuglodont - a primitive whale.

Although Dutch zoologist Bernard Heuvelmans gave literary birth to its modern form, and Costello subsequently promoted it as the last word for most lake and sea monster identities, no one appears to have taken the time nor the trouble to pursue the dynamics and biological possibility behind such a fascinating animal's implied existence in more detail. I have found this uniquely frustrating, and have therefore set out to redress this situation within these pages as best I can, the result being this possible folly of a work.

Unfortunately, I am no Heuvelmans and have had to tackle the subject without the benefit - or hindrance - of a scientific background, and have striven to undertaken the task by reviewing, expanding, and rethinking previous concepts, while offering what I feel are fresh, credible, possibilities into the theory as a whole.

The resulting speculation is not intended to convince the reader that such creatures *do* in fact exist, but is intended more to offer a thoughtful overview of the whole subject in general offering an exploration into aspects of the theory that *have not* been covered before.

It is, of course, possible that such creatures do *not* exist, but even so I feel the whole charming idea is worth a major investigation in its own right, and besides, until definite proof of Sea Serpents is offered in one convincing form or another, the case is still open in much the same way as it is for the plesiosaur and zeuglodon.

Before we begin however, there are certain things that I would like the reader to understand as to how I have presented the possible evidence.

The reader familiar with cryptozoology will notice that I have been deliberately selective in those witness accounts that I have included, concentrating on only those that seem to describe a seal-like animal or mammalian characteristics. I make no apology for this, as I wish to concentrate on

Leopard Seal (S. Lundgren)

only one type of unknown animal, and not the dozens of others that may still be `out there`, awaiting discovery. Similarly I have not dwelt too long on Lake Monsters *per se*, as I feel that there are other animals both known *and* unknown involved in such reports. Instead, I have only included a small selection to highlight similarities and to promote the idea of inland traversal rather than direct habitation for such creatures.

Along the way I have also deliberately avoided direct confrontation between mammal (seal), versus reptile (plesiosaur). This is not literary cowardice on my part, but is simply due to the fact that I am neither a palaeontologist nor zoologist, and therefore cannot comment on how good a candidate the plesiosaur actually is for the traditional sea serpent from the point of view of aquatic performance. I will leave the literal `bones of contention` (was it too stiff necked and bodied to swim as sea serpents apparently do, etc.) to others. Likewise, I have also avoided getting too embroiled in the deeper cultural and mythological beliefs that know doubt shape opinion in such context.

Finally I would like the reader to understand that although I would love such unlikely animals to be roaming the world's seas, I have tried to be as open minded and as logical as possible in my judgement regarding the evidence, and speculation.

Enjoy

Robert Cornes, 1999-2006

N.B. The original incarnation of this work was my website: The Surreal Seal Campaign. Completed 2000 but not updated since.

INTRODUCTION

The study of cryptozoology, into which this work rather awkwardly fits, has always been a fascinating and inspiring one. The speculation that there are still large, unknown animal species roaming the decreasing vastness of our planet, provides the fascination, while the discovery of the coelacanth, okapi, megamouth shark and more recently the Vu Quang ox and others, provides the inspiration.

Perhaps nowhere is this inspiration greater than in the search for the elusive sea serpent. This enigmatic denizen of the deep has plenty of room to slumber, and more than enough room to sleep, and it is somewhere within this domain that the concept of a seal which has evolved a long neck belongs.

The theory, which in its modern form has been around since the late 1950s, is both intriguing and original, seemingly offering a mammalian alternative to that most popular culprit, the plesiosaur. However, despite some witness accounts that seem to allude to such an animal, more than fifty years later, like the others of its ilk, it still remains conveniently and strangely elusive.

This elusiveness, when combined with what is known about the biology and behaviour of presently known pinnipeds or seals, evokes a common sense dismissal of such an animal's existence. For how, when so many seal species are known could such a long necked, conspicuous variant, which presumably shares similar characteristics, remain unknown?

It is this nagging question on which the present work will seek to speculate, as amongst the many accounts of long necked sea serpents, there are those that seem to describe *just* such a creature. This implies, that if these accounts are genuine, then somehow, against all the odds, such creatures do, or have existed.

The aim of this work then is to accept the included reports as truthful, and most of the time at face value, in an attempt at trying to build a speculative framework in which such a creature *could* exist. Alternative identities and explanations will at times be offered and speculated upon for the purpose of open-minded debate.

Ultimately of course, it may all turn out to be a charming delusion.

Along the way however, the reader may begin to appreciate, as I have, the complexity and adaptions of the pinniped kingdom. Lurking here, among the circus acts, there are indeed some outright monsters. There is the mammoth lurching monstrosity of the male elephant seal, willing to crush anything in its path *including* its young, in search of sexual gratification, while the violent snapping jaws of the leopard seal, quite able to kill a man, may dispel any notions of `cute` and `cuddly` animals. In contrast there is the rare obscurity of the monk seals, the surrealness of the walrus and hooded seal, and of course the crowd-pleasing antics of the cavorting, playful seal lion.

Pinnipeds all, yet sometimes strikingly different in their appearance and nature. And it is within this variance, that we may begin to find some of the answers to the questions that need to be asked.

PART ONE

BACKGROUND

1. A BRIEF HISTORY OF
THE LONG NECKED SEAL

" A Long Necked Seal by God!,
A Long Necked Seal".
(author)

The notion of a seal with a long neck is not a new one.

It was originally proposed by Dr. A. C. Ouedemans Jnr. (1858-1943), a Dutch naturalist as an attempt at explaining the perplexing enigma of the `sea serpent` in general.

The popularity and identity of such a wondrous leviathan having divided both the public and scientific communities at the time.

Although he initially favoured the then, newly discovered Zeuglodon, or primitive whale as a likely candidate, Ouedemans eventually settled for a type of unknown, gargantuan seal. A seal with both a long neck and tail and one that to all intents and purposes, resembled a plesiosaur. His conclusions were based on the scientific knowledge available at the time and his own imaginative but limited interpretation of the evidence.

This reasoning took the form of a creature that he proposed was a missing link between the cetacea (whales) and pinnipeds (seals), a notion that would not be accepted by science today.

The beast he formerly created became known as `*Megophias*` or `big snake`, a name that had previously been cited for the sea serpent by an earlier naturalist, Constantin Rafinesque-Schmaltz.

Fig. 1 `Megophias` after Ouedemans.

Megophias was a mammoth beast, a true combination of seal and plesiosaur, conveniently covering all the characteristics and features of the sea serpent, representing the definitive cosmopolitan culprit. Both revered and despised by science, it subsequently managed to paddle its way around the globe, confusing both scientists and laymen alike with its striking reptilian appearance and prehistoric profile.

While in 1930, in between theorising about giant newts, Commander Rupert T. Gould mused wisely on the possibility of such an animal in, *The Case For The Sea Serpent*, it took zoologist Bernard Heuvelmans (1916-2001), to give it life in the late 1950s, albeit in a much revised form.

Heuvelmans realised that Ouedemans was probably right in such a mammalian identity for one form of sea serpent, and with the benefit of twentieth century scientific discovery and a wealth of corresponding accounts, came to a similar but more expanded view.

Heuvelmans creature however, had lost its tail and much of its size; while its nostrils had centrally migrated to the top of its head, resulting in two quaint, snorkel like breathing tubes. Unlike Ouedemans though, Heuvelmans realised that the sea serpent was not just one animal, but a menagerie of both known and unknown ones, and decided to set out to prove it.

In his definitive and painstakingly researched work, *In The Wake Of The Sea Serpents*, he analysed 587 reports of apparently unknown marine animals, dismissing 229 of them as hoaxes or mistaken identity, leaving him with 328 accounts that he considered valid descriptions. His methodology has at times been disputed, and is perhaps faulty in certain respects, but it has gone on to shape the science of cryptozoology to the present day.

From the accounts that he studied, he deduced that there must be at least seven, as of yet, unknown forms of animal that could explain them, two of which, `The Merhorse` and `Long Neck`, were species of pinniped or seal.

Fig. 2 `The Merhorse`, after Heuvelmans.

`The Merhorse`, he proposed, was a large pinniped that had evolved a purely aquatic way of life and was characterised by its very big eyes, `horse`-like head and long, flowing mane. It inhabited the twilight depths of the world's oceans and was only rarely glimpsed at the surface by man revealing itself when it did so to be the archetypal `Sea Horse` of legend.

The `Long Neck`, on the other hand, was a giant sea lion, which Hevelmans believed, was ecologically on the rise and truly cosmopolitan in its range worldwide. It had managed to evolve two breathing tubes for a more aquatic existence and was characterised by its long, flexible neck.

Both animals were giants of the pinniped kingdom, the merhorse attaining a length of between 30-100ft. while the long neck managing perhaps a more conservative 15-65ft.

Subsequent authors have suggested that these two animals, if they existed, could be of the same species; the notable differences between them being due to their differing sexual characteristics, or sexual dimorphism as it is scientifically known. Whatever the truth, the long necked seal had arrived and went on to enjoy a brief period of popularity that it has not enjoyed since.

Fig. 3 The `Long Neck` after Heuvelmans.

This popularity culminated in the mid 1970s with the publishing of Peter Costello's enchanting, *In Search Of Lake Monsters*, in which he shamelessly promoted it as the definitive culprit behind all lake and sea monster accounts. This time however Heuvelmans beasts had merged, with a loss of snorkels but a sprouting of ears.

Costello, inspired by Heuvelmans notion that the `long neck` and `merhorse` could be responsible for some typical lake monster and sea serpent sightings, went about gathering a comprehensive selection of such reports from around the world, bringing to light much information on the subject and formulating a blue print for future researchers as he did so.

Sadly though, circumstance conspired against him, for no sooner had his book been published than `amazing` underwater photographs were obtained from Loch Ness, rekindling the possibility of plesiosaur existence along with the new discoveries which were being made about the creature and its kin. As a result after this time, the theory went into something of a decline, only resurfacing briefly in various books out of respect for Heuvelmans.

Fig. 4 Costello's beast.

Homage to the idea was briefly made by Professor Roy Mackal in the mid 1980s, in his book, *Searching For Hidden Animals*, but in general the notion became unfashionable. Today, even in cryptozoological circles it remains unpopular.

Until I hope, now.

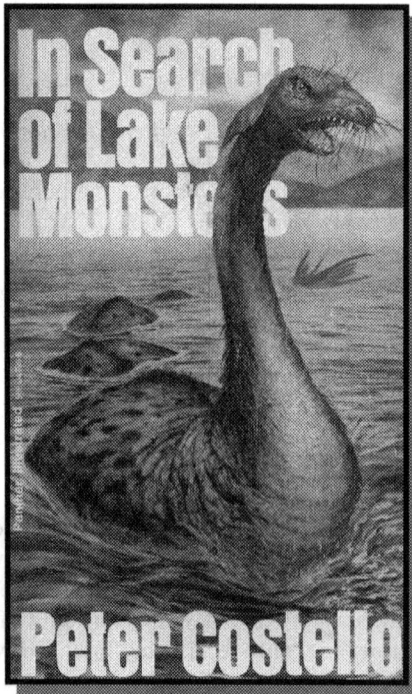

Fig. 5 Costello's book.

2. THE PINNIPEDS

- ## Why a Long Necked Seal?

Before we can move on to examine some sea serpent accounts, it may be wise to ask why we are looking at such an outlandish notion to explain at least some of them. For how, even within the wondrous artistry of nature, could such an unlikely creature such as a seal with a long neck, exist?

There are no pictures, no TV documentaries, and there is no telling carcass.

So why *not* a plesiosaur? At least we know that such reptiles at one time did frolic in the world's oceans, and if alive today, would tally remarkably well with many eyewitness reports of so-called sea serpents.

Or would they?

The fact of the matter is, that some of the most clear and concise accounts of long necked sea serpents typically identify some very mammalian and seal-like characteristics. There are, for instance, descriptions of fur, ears and even whiskers, something that no relict plesiosaur appears to have had. What is more, if we can accept these reports as genuine observations, then it follows that other less descriptive accounts must relate to a similar type of animal.

Furthermore the only existing animals adapted to an aquatic environment that have fur, whiskers and may approach the size of apparent sea serpents seals and sea lions, which with the addition of a long necked family member would at least mimic such reported creatures.

Unfortunately at present all the evidence for this notion is circumstantial. Therefore any case we can hope to build can only be theoretical and speculative in nature and to theorise and speculate, we must first take a look at the group of animals that such a creature would represent, the pinnipeds or seals.

- ## The Pinniped Family

The term pinniped, or `fin` or `feather footed`, is used to describe a group of amphibious, mammalian, aquatic carnivores that all share similarities in body shape, ecology and behaviour. Although further subdivisions can be made, they are usually divided into three main families, the Phocidae (true seals), the Otariidae (fur seals and sea lions), and the Odobenidae (walrus).

• Pinniped Evolution

At present there are two main theories of pinniped evolution.

The first, a monophyletic view, holds that all pinnipeds derived from a common, bear like ancestor in the late Oligocene or early Miocene (25 million years ago), with a subsequent divergence into the presently known families.

The second, diphyletic view, is that phocids and otariids originated independently of each other some 20 million years ago, the otariidae and odobenidae evolving from a bear like ancestor and the phocidae from otter like animals in the mid Miocene. However, with the advent of modern genetic testing, studies have shown that at a molecular level at least, phocids and otariids resemble each other more closely than any other carnivores, thus implying the monophyletic view.

The ancestors of modern pinnipeds, the pinnipedimorphs , first appeared 25-27 million years ago and although primitive they can easily be identified as being the fore runners of modern day pinnipeds sharing the same basic shape and ecology of today's species.

It is generally thought that the fossil record of pinnipeds reveals only a small fraction of what was once a greater diversity among the order and there are many palaentological gaps especially within the otariidae. It is thought that all existing otariids have evolved in the last 2-3 million years while the differentiation between the phocids and otariids has occurred in the last 2-5 million years.

Fig. 6 Enaliarctos mealsi. (Pinnipedimorph)
(estimated length 1.4-1.5 metres).

Despite this possibility of common ancestry, there are still some distinct differences between the modern families, some of which are summarised below.

• The Phocidae

Phocid seals are sometimes referred to as `earless seals`, as they lack noticeable external ears, (although some grey seals may in fact grow small external ear pinnae). Instead they have two openings into the ear canal, situated behind their eyes. Their fore flippers are short while their hind ones are longer, and all of the flippers are clawed.

Fig. 7 Crabeater Seal (M. Cameron)
National Marine Mammal Laboratory

When swimming, phocids use a 'sculling' action of their hindquarters and hind flippers, the fore flippers being used to steer at slow speeds.

On land they move awkwardly, the hind flippers useless for locomotion, and although some species may pull themselves along using their fore flippers, they tend to lurch or bounce ungracefully and clumsily along the ground.

Reproductive behaviour in phocids varies, but compared to the other two families they have a shorter period of pup weaning and dependency.

Because of this and their poor locomotion on land, phocids are usually considered to be the most advanced of the pinnipeds, gradually evolving to a purely aquatic existence in the same way that dolphins and whales must have done.

On the whole they tend to be larger than otariids and more streamlined. They are able to dive deeper and longer than the other two families.

• The Otariidae

Their fore flippers are long and have splayed digits, the first digit being longer and stronger than the rest. They have small nails situated well away from the flipper edge and these are also present in their hind flippers, along with further rudimentary claws on digits one and five.

Fig. 8 Steller's Sea Lions (NMML).

Their fore flippers are long and have splayed digits, the first digit being longer and stronger than the rest. They have small nails situated well away from the flipper edge and these are also present in their hind flippers, along with further rudimentary claws on digits one and five.

Their waterproof pelage is generally thicker than phocids and they tend to have less blubber.

In the water otariids extend their head and neck for steering and swim with a strong, breaststroke action, the hind flippers being used as stabilisers. On land, they can rotate their hind flippers forward, using them to walk or bound, lifting their bodies clear of the ground. At the same time they swing their head and neck from side to side creating momentum for their movement.

All otariids share the same polygamous reproductive behaviour to a certain extent and compared to phocids, they have a longer pup weaning and pup dependency period. Although sexual dimorphism is present in both otariids and phocids, it is usually more pronounced and striking in otariids. For instance mature male fur seals have visible ruffled manes.

• The Odobenidae

The odobenids are unusual, as any picture will show, in the fact that although they form a separate family order, they share both phocid and otariid characteristics. Like phocids, they do not possess obvious external ears while like otariids they can use their hind flippers for locomotion on land. When they swim, they use a mixture of both phocid and otariid technique. At one time they were thought to be more closely related to the otariids but are now thought to be closer to the phocids.

All in all, they are a bit of an odd mix.

The odobenids of course, also have a unique attribute of their own and one that is not found in the other two families, their tusks. These unwieldy assets are used primarily in sexual display and for hauling onto and digging into the ice as well as feeding. Although present today, fossil odobenids do not appear to have had them. Fossil evidence has also indicated that in the past there were around thirteen different species and early walruses called dusignathines, resembled sea lions.

Of all the pinnipeds, they have the longest pup weaning and dependency, and are also the slowest swimmers.

• Distribution

There are thought to be some 50 million pinnipeds in existence today, 90 percent being phocid, the remaining ten percent otariids and odobenids.

It is thought that phocids first appeared in the North Atlantic about 15 million years ago, their ancestors having migrated through The Central American Seaway, and following deteriorating climatic conditions it is thought that they migrated both north and south.

The otariids first appear in the North Pacific, approximately 10 million years ago, from where they migrated southwards, either through The Central American Seaway, or along the coast of South America (or both), with the sea lion family diverging about 3 million years ago to form a separate family.

The Odobenids first appeared in the North Pacific 5-8 million years ago, where it is thought that they migrated first into the North Atlantic before returning to the Pacific via the Arctic.

Today, some species such as the Guadalupe fur seal and Mediterranean monk seal are confined to specific areas, while species such as the elephant seals, occupy large ranges in both hemispheres.

Most pinnipeds tend to migrate and wander widely from their habitats, sometimes travelling vast distances and many species overlap each other in their distribution patterns. Walruses are found in both the North Atlantic and North Pacific Arctic regions while most fur seals are found in the southern hemisphere. Sea lions are found in both.
No otariid species inhabits the extreme Polar Regions or North Atlantic, while no species of pinniped is native to the Indian Ocean.

- ## Adaptions

During their evolution pinnipeds have become well adapted to their particular ecological niche and in recent years, with the outlawing of massive seal culls, many species have begun to increase in number and recover from previous depletion. They are intelligent, efficient and versatile animals able to adapt various strategies for survival sometimes in some very inhospitable conditions. Life expectancy on average can be 20-30 years although the oldest pinniped in the world, (documented from the Shetland Isles), made the ripe old age of 46.[1]

Some species such as the Baikal seal, a lake-living pinniped from Russia, have adapted a freshwater environment to live in although all pinnipeds can live in both fresh and marine water.

As well as being tremendous swimmers, pinnipeds are also able to dive to great depths, sometimes for quite considerable periods of time. Some like the Weddell seal have been recorded to dive for over an hour (one recorded dive took 73 minutes). How pinnipeds manage this is still something of a mystery.

It is not thought that they use echolocation as some cetaceans do, but there is strong evidence to suggest that they use a similar sound locating method.

- ## Characteristics

Most pinnipeds usually exhibit playful, non-aggressive behaviour and can be shy. Being mammals they are often also curious in their nature. Most are sociable interacting well with others of their species and in certain cases, such as the Californian sea lion, humans as well.

There are of course exceptions to this rule as the following pages will show and the public perception of the behaviour and appearance of pinnipeds tends to be shaped by their knowledge of familiar seals observed through various popular mediums. Such demeanour is lacking in some species such as the unpredictable leopard seal while others like the Mediterranean monk seal are extremely private and reclusive creatures, deliberately shying away from human attention.

Despite their many adaptions, present-day pinnipeds, unlike the ceteceans, have not evolved away from their need to rely on land for certain stages of their lives and all presently known pinnipeds must return to land to give birth.

Therefore, if a long necked variant exists, it must have a similar need, further implying that if it did exist, it would have been noticed somewhere by now.

Well, as the following pages will show, maybe it has.

PART TWO

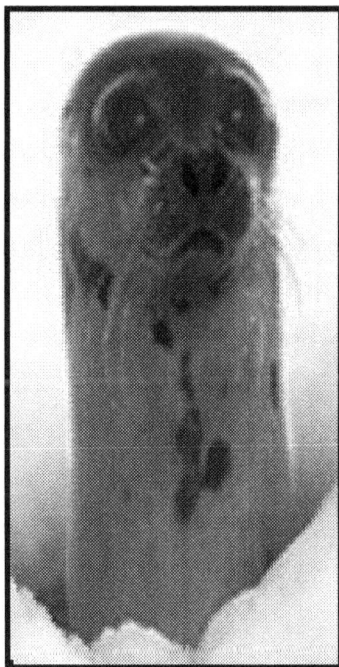

THE EVIDENCE?

Tantalising accounts from around the world.....

3. VAGUELY FAMILIAR?

In the following pages I have included various accounts of so called 'sea serpents' from around the world all of which seem to describe at least some typical pinniped or mammalian characteristics. This makes them both relevant to this work and thus vaguely familiar.

Sometimes these descriptions are so seal like that the most unusual aspect of them is why they were not accepted as such animals rather than be classed 'sea serpents' or 'unknowns' in the first place.

It would appear however that there is something that always seems to separate them from this identity in the eyes of the witnesses. Whether this is due to a combination of surprise, wishful thinking or the lack of knowledge on the part of the observer it makes the accounts extremely frustrating in nature, always hinting but never clearly defining.

- **Delusion Down Under?**

The first account I have included is both and interesting and useful one. It is interesting in the fact that it dates from 1857, a relatively plesiosaur free time and is useful in the fact that it allows us to develop a line of reasoning to exclude known animal species. This reasoning can then be adapted and applied in some instances to future pages.

It originates from South East Australia at a time when the white settlers were becoming established and had began to explore the vast territories that existed, learning along the way of the rich sometimes confusing, aboriginal culture.

Part of this culture was the belief in an unusual form of freshwater denizen known culturally as the 'bunyip', a strange and enigmatic creature, which appeared to be able to take on several different forms. One of these forms was a short-necked semi aquatic creature, attributed with various characteristics and variously described, said to inhabit outback billabongs and remote stretches of water where it seems to have been greatly feared. Another form appears to have been a long necked creature, combining a number of native animal components in its description. This bunyip was found predominantly in the rivers and lakes of South East Australia and it is the composite nature of this beast that is of interest to us here.

The report is a typical 'white man's bunyip' encounter as described by a Mr Edwin Stoqueler, an artist come naturalist, who took it upon himself to sail down the Murray and Goulburn rivers in a canvas boat. He apparently spent much of 1856/7 doing just this, whiling away the time compiling sketches for a diorama he was working on. When completed this was then apparently to be published in England. The account was reported in the *Moreton Bay Free Press* of 15/4/1857 and can be found in Malcolm Smith's excellent 'cryptozoology down under' book, *Bunyips and Bigfoots*, as well as

Costello's work. Thanks to Pam Cory of the Brisbane Historic Society, I have been able to view a copy of the original newspaper report myself.[2]

> *Mr Stoqueler informs us that the Bunyip is a large freshwater seal having two small paddles or fins attached to the shoulders, a long swan like neck, a head like a dog, and a curious bag hanging under the jaw, resembling the pouch of a pelican. The animal is covered with hair like the platypus, and the colour is a glossy black. Mr Stoqueler saw no less than six of the curious animals at different times, his boat was within 30ft. of one, near M'Guires point, on the Goulburn and fired at the Bunyip, but did not succeed in capturing him. The smallest appeared to be about 5 ft. in length, and the largest exceeded 15 ft. The head of the largest was the size of a bullocks head and 3 ft. out of the water.* *

Had it not been for the lack of a second barrel on his gun, and the flimsy nature of his boat, Mr Stoqueler would have apparently made a serious attempt at killing one of the creatures.

Now, apart from the `swan`-like necks of the creatures, which Smith rightly points out are not easily reconciled with the 3 ft. mentioned, it would appear that Stoqueler was observing seals although it is not clear if the paper nominated them `freshwater seals`, or whether Stoqueler did.

Although many species of pinniped can wander widely, the native pinnipeds of Australia , are the Australian sea lion (*Neophoca cinerea*), discovered in 1816, and the South African or Australian fur seal (1776, *Arctocephalus pusillus*). †

Although in their neighbour, the New Zealand or Hookers sea lion (1844, *Phocarctus hookeri*), adult males may reach 2-3.25 metres in length (6-10ft.), the males of both these Australian species average 2-2.5 metres (6-8 ft.), while the females of all the species are smaller. So even accounting for such a poor judgement of length at such a close distance, neither reaches the 15ft. plus of the largest animal seen but average lengths can of course be exceeded.

---

* *It is beyond the scope of this present work to get embroiled in further analysis and differentiation between the various types of bunyip but a brief glance at the bibliography section will give ideas for further reading. A typical bunyip described by the aboriginal people of the Port Philip district, `Tunatpan`, is described as aquatic with an elongated neck and head resembling an emu, a mane like a horse, flippers like a seal and a horse like tail. `Too-roo-don`, a bunyip of the North West Victoria aboriginal people is similar. In fact these peoples differentiated between two types, `banib`, pronounced bunnip, a lake living pig like creature and `banip-ba-gunuwar` meaning bunyip and swan[1]*

† *The Australian fur seal, described in 1904 has slight cranial differences to the South African fur seal but is classed as the same species.*

Although otariids have flexible necks that can appear quite long when extended, Stoqueler likens them to the necks of swans, which are thin and distinct. While bunyip reports have mainly been equated with seal like animals over the years, an otter like identity has also at times been suggested.

An otter would have a `glossy` coat and its outstretched neck may appear thinner and more distinct than a seals and in fact the South American giant otter can reach well over 6 ft. But a fifteen-foot otter is presently unknown. However, although Stoqueler is close enough to report `hair`, he makes no mention of ears or whiskers which otters and otariids both have, (ears sleeked back may not be very noticeable in otariids), while he does compare the heads to those of dogs which do have ears. What is more, he is close enough to report another curious feature, a `pouch like that of a `pelican`, (although coming from the land of marsupials, I suppose we should expect this), a most unusual and intriguing one.

No pinniped or otter displays such a feature, unless the pouch was merely folded skin. The nearest possible comparison that could perhaps be made with a pinniped, is the walrus, which has two pharyngeal pouches either side of the oesophagus which can be inflated to hold up to 13 gallons of air[3]. These are not however visible. This attribute is found in both sexes, but is more developed in the male and is used variously as a buoyancy aid allowing the animal to float, while also enabling it to make specific sounds, which are of importance in mating. Walruses are also obviously quite distinct and do not inhabit the southern hemisphere.

So although going some way to explain what Stoqueler saw, neither an otariid or otter fits the whole description, meaning that we must take a look at some other possibilities.

There are two other possible pinniped contenders, but again neither fits the bill completely. These are the leopard seal (1820, Hydrurga *leptonyx)*, and the southern elephant seal (1758, *Mirounga leonina*), both usually Antarctic dwellers but known to roam widely. The elephant seal is a veritable mammoth among pinnipeds, reaching up to and over 5 metres (approx. 20ft.) in the male of the species and the male does have a sort of pouch, its inflatable proboscis, although this is situated on the front and top of its head.

The leopard seal, which can reach 3 metres (10ft.), is in fact renown for cutting a sinuous figure and has a neck and head to match, however its pelage is silvery grey to blue. Both animals are known to wander from their regional climes and are phocids, which would mean they would not have ears. In fact, according to T. Healey and P. Cropper in another `aussie` crypto book, "*Out of the Shadows*", elephant seals were once quite common in some parts of Australia, while a leopard seal was captured 48 km. up the Shoalhaven River in 1870, complete with a platypus in its stomach.

Fig. 9 Elephant Seals. (Karen French)

In both cases however, a distinct swan like neck would be hard to demonstrate. Leopard seals are also mainly loners and aggressively curious in nature, so although it would not be impossible for leopard seals to be found so far from home, even accounting for the possibility that Stoqueler saw one or more of the animals twice, it would be unlikely.

If we also take into account the fact that Stoqueler took a shot at one, it would have been unusual for the animal, if a leopard seal, not to have elicited some response. Besides all this, Stoqueler, having spent some considerable time witnessing the local fauna, presumably becoming familiar with certain pinniped species, would have surely recognised them for what they really were, known seals.If he had also spent much of 1856/7 sailing up and down these stretches of water it might be expected that he would have in fact made sightings of wandering seals imitating bunyips more frequently, allowing him to distinguish clearly between the two. By all accounts he appears to have been regarded as a reliable witness, running for a position in the local police force at the time of his encounter. He made a sketch of one of the animals, which was then viewed by some of the local aborigines. They informed him that the drawing showed the bunyip's brother and this was interpreted by the paper as meaning an exact likeness of the bunyip.

Unfortunately the sketch does not appear to have withstood the passing of time, but apparently showed the head and neck from one of the freshwater bunyips. This is a shame as if he were an artist of some talent we may have probably gained a very accurate picture of the creature.

A reading of many `bunyip` accounts by various researchers and the fact that seals have been positively identified around inland Australia indicates that many such sightings are simply due to `out of the way` pinnipeds. A group of seals for instance, was observed 1200km from the sea, swimming up the Murrumbridge river in 1850, while a pinniped was apparently shot and mounted in the same year 1500km from the ocean in New South Wales. One was also observed 400km up the Murray River itself in 1890 [4] while in 1947 four seals were reported to be inhabiting the Mulwaree river, about 5 miles from Goulburn.

It has also been noted that modern bunyip reports are few and far between and may indicate that people today are more familiar with seals, which were previously mistaken for them. There are also likely to be many more man made obstacles today that may prohibit wandering creatures such as seals to travel inland.

Ten years previously, another bunyip had been seen, this time at a cattle station by the junction of the Lachlan and Murrumbridge Rivers after extensive flooding had taken place. The whole encounter was reported in the *Argus* on the 29/6/1847. This time the bunyip had been found, `grazing`! It was:

> " *as big as a six month calf, dark brown colour, long neck and long pointed head, with a thick mane of hair from its head down to its neck and large ears that pricked up. It had a shambolling gallop and a large tail with its fore quarters large in proportion to its hind quarters.*" *

Both man and beast fled from the scene of the encounter.

However, instead of a pelicans pouch, this creature had two `tusks` no less! A sabre toothed sea lion?!, Heuvelman's snorkels?! We will never know.

Still around Australia, we will next take a look at an often-quoted account of a sea serpent, seen on land. This encounter happened in Tasmania, in 1913, April again. It was related by a mining engineer prospecting on the west coast of Tasmania who had learned of it from the two men, his colleagues, who had witnessed it.

Oscar Davies and his mate, W. Harris, who were apparently quite familiar with `sea leopards`, seals and sea lions, were walking along the coast, when they noticed a dark object at a distance of about half a mile. It was under some sand dunes and appeared to be showing some signs of movement.

They advanced to within about 50 yards of it, when;

> " *It rose suddenly and rushed down to the sea. After going about 30 yards, it stopped, turned around showing only the head for about 5 seconds and then disappeared under the water*".

---

* *Aboriginal people of the Murrumbridge area had a firm belief in `katenpai`, `kinepratia` and tanatbah according to dialect. This animal could grow as big as a bullock, again with an emu like head and neck, mane, horse like tail. It also had four legs with three flipper like webbed toes on each foot.* [5]

They described it thus:

> " It was 15 ft. long, it had a very small head, only about the size
> of a Kangaroo dog. It had a thick arched neck, passing gradually
> into the barrel of the body. It had no definite tail, and no fins. It
> was furred, the coat in appearance resembling that of a horse of
> chestnut colour, well groomed and shining. It had four distinct
> legs. It travelled by bounding, that is arching its back and gather-
> ing up its body so that the footprints of the fore feet were level
> with those of the hind feet. It made definite footprints. These
> showed circular impressions with a diameter (measured) of 9
> inches and the marks of claws, about 7 inches long, extending
> outward from the body. There was no evidence for or against
> webbing.
>
> The creature travelled very fast. A Kangaroo dog followed it
> hard on course to the water and in that distance gained about 30
> ft. When disturbed it reared up and turned on its hind legs, its
> height when standing on four legs would be 3ft.6 - 4ft. "

Firstly, although the men were apparently familiar with pinnipeds, they do not seem to associate them with this remarkably seal like animal. The creature itself, is not very big as sea serpents go and as both Smith and Heuvelmans point out, no mention is made of the familiar long neck, but instead, a `neck that passed *gradually* into the barrel of the body`. However, this cannot have been very gradual if we take into account the animals height of 3-4ft.

So what did the men see?

It isn't an otter, as apart from the size, there is no conspicuous tail and a phocid seems to be out of the running literally, as the animals movements sound very otariid in nature and unless the dog was deliberately trying to avoid it, the creature must have been moving at a fair old clip.

Although the animal is described as having four distinct legs, it could be supposed that if the beach was sandy and the animal heavy, a fair amount of sand may have been dis-placed as it made its escape, which possibly obscured the true nature and shape of its limbs. The men cannot have been too sure about these as they state there was no evi-dence "for or against webbing" possibly indicating that they did not get that good a look at them. Definite footprints were made though, with the apparent impressions of claws, but it is not clear whether the claws were present in all the footprints, which may have helped identity the creature. All in all is tempting to argue that all the men saw was maybe a pinniped that they were not familiar with (although from their statement this cannot have been many), a possible example perhaps being a fur seal whose fur had dried out giving it an unfamiliar appearance. Even so, the encounter if not of some type of seal, must remain a true, baffling 'unknown', as it should be remembered that the men had a pretty good view, walking from a distance of half a mile to within 50 yards of it, presumably keeping it in view for the whole time.

All very frustrating.

Heuvelmans concluded that the men had seen a juvenile 'long neck', however all I think that can possibly be concluded is that they saw a very seal like animal, with the ensuing debacle obscuring its true nature.

- ## Mammal or Reptile?

One of the most argued points with regard to the traditional long necked sea serpent is the question of whether such creatures if they exist are mammals or reptiles. The above three reports which describe fur must obviously relate mammals whether they are known animals or a new species. The greater debate will inevitably continue until physical evidence of one sort or other is definitively obtained and presented for every-one to see.

In the following pages I have included reports of animals that again sound more mammal than reptile but will endeavour not to take sides but merely present *some* of the evidence that I feel indicates a mammal in these accounts. Although classed as sea serpent accounts, I have tried to explore more mundane possibilities where possible.

- ## Of Horses, Manes and Men

In October 1997, a Mr Pickard was doing a spot of night fishing on a beach at Alde-burgh in Suffolk, England [6]. It was about 2 a.m. and he was not enjoying much luck. He moved further down the beach to cast off when he saw what he first took for a tree trunk floating in the water. It appeared to have a tangled looking mass of exposed roots at one end. However after it began bobbing up and down in the water he realised that it was not merely a piece of wood. Moreover the root like projections appeared to be some form of horny growth on a rather unusual animal that was swimming off shore. He could see 8-10ft. of this creature, which subsequently changed course and began swimming inland. As he watched he became convinced that he was observing a horse, possibly washed overboard during recent storms in the area.

He was just about to phone the RSPCA to report this matter when the creature turned parallel to the shore and he then realised that it was not a horse, but a very unusual animal the like of which he had never seen before. As he watched, the animal proceeded to put its ears back at a 45-degree angle and disappear beneath the water.

Following his encounter he tried to find an explanation for what he had seen and in a round about way came across my website. He emailed me through a work colleague, giving his telephone number, and I subsequently phoned him and discussed things further. Having come across some cryptozoological references he was convinced that he had seen one of Heuvelmans' merhorses.

Now this is quite an intriguing encounter and if *not* a horse swimming (underwater!?) or similar terrestrial animal then it is hard to determine exactly what he saw. He was extremely honest in his account and did not claim to have seen a long neck or any other common preconceptions of what a sea serpent might look like.

Logically you would expect any animal that displays external ears to be a mammal and it would be tempting to argue that what he saw was a seal, as certainly in profile it does seem to bear a resemblance. However, no native seal species of the British Isles has visible ears (an exception of course could be a harbour seal that had developed pinnae *or* an alien species). Interestingly, in profile, the ears appear quite short compared to the front view. There had been a couple of reports of unusual and unidentified animals from around this area in previous years but specific detail from them is lacking. Sceptics no doubt, will argue that this *is* merely a report of a horse that *had* decided, for whatever reason, to take a late night dip.

This will not be the last time we encounter mammalian features such as ears, as they will literally `crop` up in future accounts. Where those accounts relate to an animal, which bares a long neck it, should be a good indication of a mammalian identity. What should be remembered however is that the accounts I have included are only a few from *many* hundreds which do not include such details. Strikingly similar creatures have been seen the world over, sometimes very closely, yet such features have not been reported in these accounts.

So does this mean that there is a whole menagerie of unknowns out there or does it imply that witnesses selectively pick out certain bits of information based on some subconscious opinion that they already have or want to perceive?

Misidentification of known animals in unfamiliar environments probably plays a part in some less detailed accounts of potential sea serpents and will always be used to explain them. Therefore to present a philosophical approach to this subject it is probably a good time to introduce some lateral thinking, with regard to alternative possible identities before we go any further. This is not designed to refute or discredit any of the included accounts but will hopefully show an open mind on my part and allow the reader speculate on all possibilities. This is important as sea serpents and lake monsters the world over do seem to take on a bewildering variety of guises.

Firstly it is well known that many species of marine sea creatures end up far from home during their travels. Seals are notorious travellers and many species can travel thousands of miles away from their normal range. Hooded seals for instance usually found in Arctic waters, have been spotted as far a field as Puerto Rico. Harp and ringed seals have also been found wandering the British coast as far as Cornwall.[7] One walrus made it to Norfolk in 1981 while another turned up on some rocks in County Mayo, Ireland

Figs. 10 Mr Picard's creature.
(from CFZ archives)

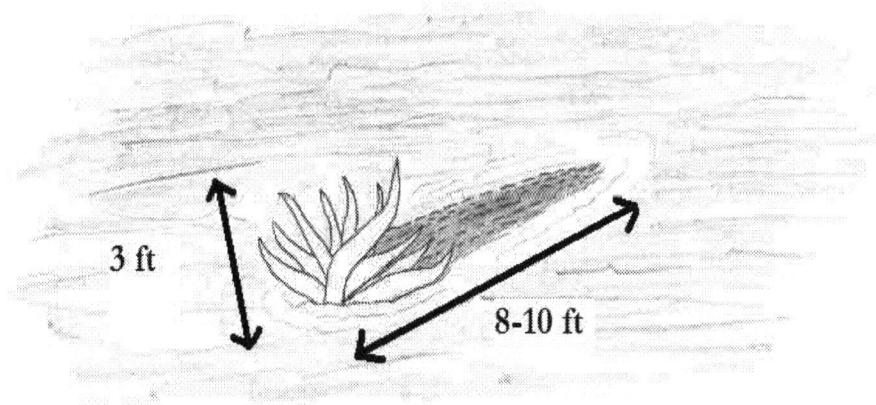

3 ft

8-10 ft

We also interviewed Mr Picard and he gave us these pictures,
complete with measurements

18 in

40-60 feet off the beach

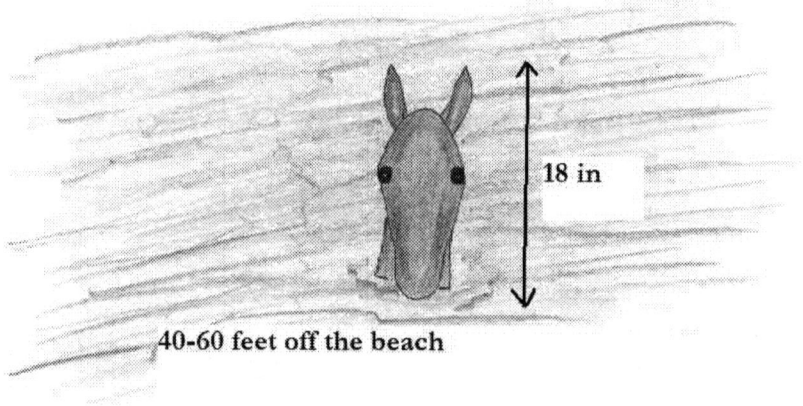

about 25 ft from shore

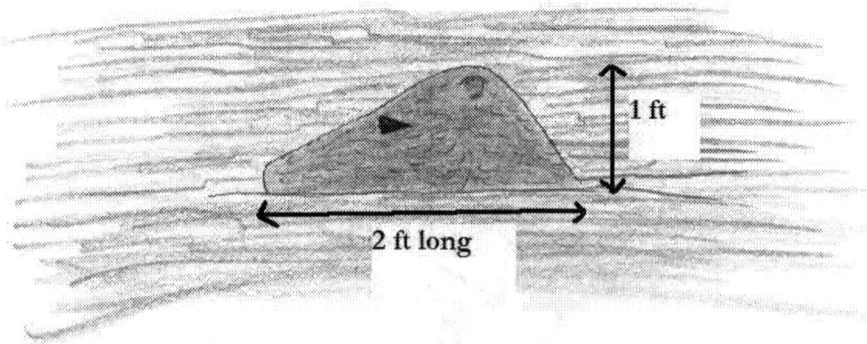

1 ft

2 ft long

in 1999. Previous to this a dead one had been found in County Kerry in 1995.[8] Recent research into their mating habits, has also shown that bull elephant seals may travel as far as 8000 miles for a good night out. In fact elephant seals may spend 3-7 months foraging for food on journeys that may take them thousands of miles from their habitat[9]. Also seals, dolphins and the like are able to swim into rivers and lakes while seals can also travel long distances over land. In fact I would suggest that many incidences of lake monsters and some sea serpent accounts are simply due to the wanderlust of these animals.

A recent survey has also estimated that there are probably up to 3000 whales and dolphins cruising the seas around the UK and when you then add porpoises, oarfish (long serpentine fish with red crests that can reach 30 feet or more), sunfish and leatherback turtles (which can reach over 5 ft. in length), again, all of which have been found in British waters, it becomes rather muddled.[10] If we now add all the possible combinations of marine life found in different parts of the world where sea serpents have also been reported we may end up totally bewildered.

More surprisingly and to add to this global confusion, it is not just marine animals that take to the water. Pigs from the Tokelau Island in the South Pacific have been known to wander over coral reefs at low tide, while domestic sheep in Scotland have been swimming out and foraging on the intertidal kelp beds for hundreds of years. Deer can swim and have been seen doing just this at Loch Ness. In fact sika deer, introduced to Britain from Japan, can apparently swim up to 12km in the sea while their cousins, moose, in Scandinavia, have been reported swimming from Sweden to Denmark. [11,12] Moose are also apparently accomplished divers and can actually dive to 4 or 5 metres and stay down for 1-2 minutes before resurfacing.

Not to be out done, there are also `udder` accounts of sea going cows trawling the coasts, milk floats indeed!

However apart from the general reported size of sea serpents, which seem to exceed these land creatures and the fact that such accounts usually relate to fast moving animals, which are seemingly at home in the water, it is not unreasonable to assume *some* possible `unknowns` are simply due to seeing a familiar land animal, *unfamiliar* in the sea. However for any land animal to *surface*, then swim, then *submerge* and disappear is quite a feat.

Moving on....

The following report is as far as I know unpublished and was sent to me by a Mrs Wilma Greenwald to whom I owe many thanks.

It came to light in defence of some of the scepticism that was being levelled at the Loch Ness affair during one of its regular flare ups in the 50s and was subsequently reported at a much later date than when it actually took place most notably in the *Aberdeen Eve-*

ning Express 19/5/1958.

It was witnessed by the father of Mrs Greenwalds uncle, a Mr Aitken (senior), who was at the time a Scottish fisheries manager and dates to somewhere between 1874-1878 on a day when he was out in the water of Loch Broom fishing.[13]

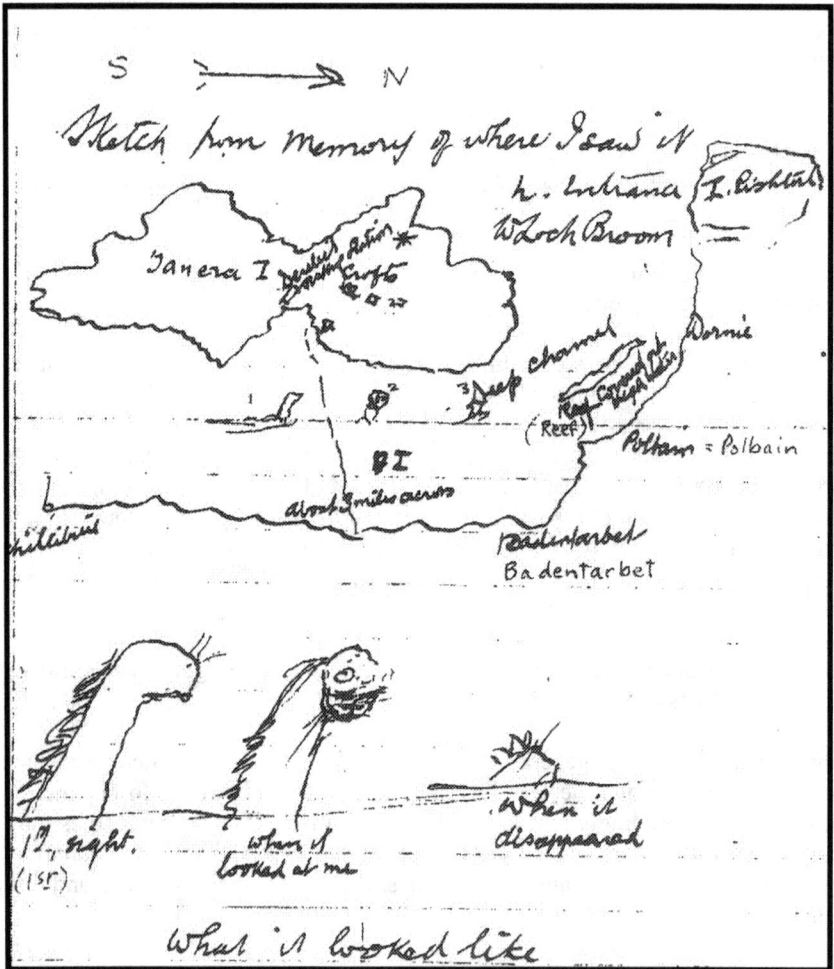

Fig. 11 Mr Aitken's sketch.

Mr Aitken was alone in a boat at the end of the loch, which is a sea loch in the North West of Scotland, when he heard a rushing sound like that of a waterfall despite the fact that the loch was perfectly calm and still. When he looked round for the source of the noise he saw, about 500 yards away:

> *".. a monster about 10 ft. high rushing along at a great rate throwing up spray and waves like a steamer. The monster turned its head and looked straight at me, then probably sensing danger from the boat, quickly submerged.*
>
> *The neck above the water was about 10 ft. high and 3ft. diameter. On the back of the neck there was a mane which looked like the tips of tangles, and it had some sea weed like growths about its mouth. It had two large glassy eyes set high on the head. The whole colour was that of dark seaweed."* *

Mr Aitken later sketched the beast along with the positioning of his boat when the sighting took place including various details regarding the animal's course while in the loch.

From the written information given to me by Mrs Greenwald I can find no reason for doubting or disputing this report. Here we have a creature with a 10ft. neck, apparent mane to boot and one that seems to be moving at fair speed.

The drawing also seems to include what appear to be whiskers (? seaweed like growths). Heuvelmans, the reader will remember, included both such features on his `merhorse`, of which this creature is strikingly reminiscent.

Taking into account the recent discussion into mistaken identity, there are few if any animals, which could give the above impression, even if we reduce the length of the neck to some degree. None that are presently known anyway.

The sketch is also interesting, as it typifies the sort of appearance that is common when describing a `mane` seen on a sea serpent, of which there are many accounts.

A very similar description to Mr Aitken's impression of a mane is given from a Mrs Hildegarde Forbes who in 1922 was a passenger on a steamer bound for Alaska The boat was passing along the east coast of Vancouver Island when she saw a forty-foot `Monster`. The creature had a snake like raised head; five to seven humps with *again* a mane that `seemed like seaweed`.

Similar such accounts have been noted the world over and figure most conspicuously in Celtic and Scandinavian folklore. These are often echoed worldwide, possibly as a result of migrant colonisation. There are also beliefs and motifs associated with such a feature, some with a deep cultural significance.

Heuvelmans in fact includes two early reports from Norway in his book, both of which

---

* *If Mr Aitken was referring to the local kelp in his description of the mane, it could be assumed that it would therefore be brown in colour.*

mention manes. One from 1837 describes an enormous creature with a horse like head and mane, a moustache like a seals and big black eyes apparently as `big as saucers`. The other more detailed account was related by a Norwegian fisherman who while practising his livelihood one day was disturbed and cogitated over by a large inquisitive animal about `five to six fathoms` long with a body `as round as a snakes`, two feet in diameter. The head was as thick and as long as ten-gallon cask with five inch glistening red eyes. The serpent held its head at an angle above the water and close behind this spreading on both sides and floating on the water, was a brown mane of `tolerably long hair`.

As intriguing as these last two accounts are, apart from the large size given to the observed creatures, they sound like some form of seal. Although we have no obvious description of a long neck in these the appearance of the manes seems to indicate that they are visibly noticeable, that is they are <u>distinct</u>. If the animals were some kind of known maned, pinnipeds with water-smoothed fur you would perhaps expect that their manes would not be as readily distinguishable. If, however, these or similar creatures represented some form of long necked pinniped which was related for instance to the otariid clan, then like mature male fur seals a mane could be present. If this was the case it could be assumed logically that the mane may give a flowing appearance, as there would be a longer neck and shoulder region for it to cover in such an animal.

Whatever the nature of this attribute it clearly stands out, as a further report from North America will illustrate, interestingly also from around the Vancouver Island region.

Two women, a Mrs Stout and Mrs Parson, accompanied by their respective young sons, were walking along Dungeness Spit in British Columbia on a rainy, foggy March day in 1961. They were watching a large freighter through the mist when they saw what they first took to be a tree limb (sound familiar?). This disappeared beneath the water before reappearing again. When it did so they saw a flattish head and three humps behind a long neck appear. The object, some type of animal was quite distinct in the poor visibility and appeared to be observing the ship. The women described it as having a `floppy` mane or `fin` on its neck and in a sketch, which was later drawn, this feature is quite strikingly illustrated.

The whole encounter lasted about 8 minutes and left them quite confused as to what they had actually witnessed. Indeed Mrs Stout who was in fact a trained biologist, could not reconcile such a feature with any known animal.

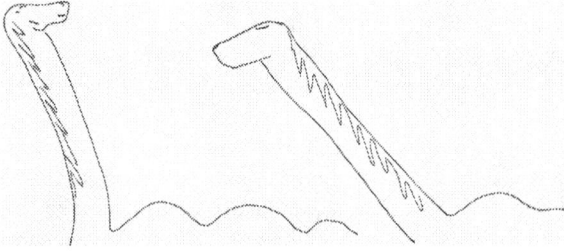

Fig.12 Two views of the Dungeness Sea Serpent.

We will find further reports from British Columbia relating to the resident sea serpent there, known as Caddy - short for Cadborosaurus - just as revealing in future pages. Before we move on, though, I will include one more report from this general area.

This sighting took place on the coast of Oregon in an area where a natural chasm breaks the rocky shoreline and is known locally as `Devils Churn`*. In 1937 a Mr and Mrs White were watching spectacular wave breaks across the `Churn` on another rainy and this time stormy day. They spotted an animal at the mouth of the Churn swimming towards the shore and were shortly joined by a passing truck driver who had also seen it appear as he was driving along a nearby road.

It had a long neck, apparently 15 foot, which reminded the witnesses variously of a horse, a giraffe or a camel. It had a mane the colour of seaweed (again), which in this locality is also brown, that was visible on the neck trailing all the way down to the body. A tail was seemingly noticed giving the impression of an animal with a length of about 55 feet. On the head were two projections which Mr Hunt took to be small fluttering ears while his wife thought that they were really 8-10 inch small, straight horns.

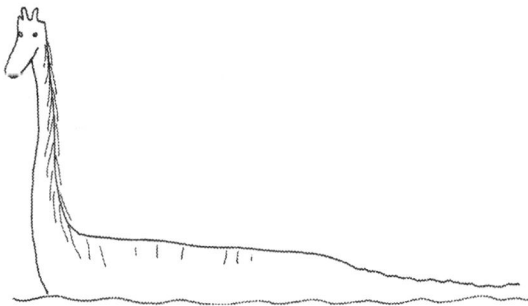

Fig.13 The Devil's Churn Sea Serpent

---

* *Coincidentally the Devils Churn it is about a dozen miles north of one of the largest underwater cave systems in the world, known as the Sea Lion Caves famous, you guessed it, for the sea lions that inhabit them.*

Although the above sketch may remind the reader of a hobbyhorse we are left with a very similar creature to the Dungeness Spit one. It is seen in an area haunted by some unknown animal, has a mane and this time also has small ear like appendages.

What appears to be the tail in the sketch looks more like the tail end to me rather than anything definite while the `ears` or `horns` seem out of proportion for such a big animal. However if we are looking for a pinniped identity and take the horns as definite ears then there is probably a strong resemblance to how they may appear on a similarly sized otariid seal.

Moving closer to home now, another eared `something` was seen off the coast of Cork (Ireland) in 1907.

The witness was the future commodore of the Cunard shipping line, Sir Arthur Rostron who later recounted the details of his unusual sighting in his memoirs.

While chief officer on board The Campania which was on its way into port, an unusual looking animal was spotted about 50 feet from the boat and was seen to be *"turning its head from side to side"*. The head itself was 8-9 feet out of the water on a 12-inch thick neck. It had *"two small protuberances where the eyes should have been"*, and very small ears. Rostron produced a sketch:

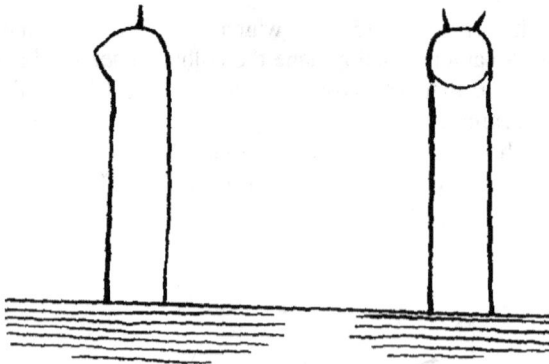

Fig 14. Rostron's sketch.

The projections in his sketch do give the impression of ears but things are slightly confused when the matters of the `small protuberances` are taken into account.

Three years later and further up the coast another sighting was made and makes for interesting reading.

It occurred in 1910 and was witnessed by a Howard St. George and one of his sons on the coast of Connemera in Mayo.

An animal was sighted floating on an ebb tide into Kilkerrin Bay. It was *" as big as a large 2-horse lorry"* with a hairy body about twenty feet long. It had a head at the end of a six-foot long neck, which again was swaying from side to side as if the creature was looking for something.

What this something may have been is anybody's guess.

Nearly twenty-five years later in 1934 back in Scotland, on the east side of Campletown Loch on the Kintyre peninsula John MacCorkindale, a local naturalist, was talking to a postman on a road, which skirted the shore. Their conversation was rudely interrupted by a loud splash. When they turned round to the source of the noise they saw an enormous creature about 300 yards from shore.

It would apparently raise the front part of its body out of the water for about 12 feet and then flop down in the water, something that it repeated several times before finally giving up the ghost, submerging and disappearing from their view. MacCorkindale a trained observer later described the animal as having a fore body similar to that of a giraffe with a long thin neck, small head and ears reminding him somewhat of a land animal. It was a glistening silvery colour and was about 30 feet long.

MacCorkindale also got the impression of some `trace of a dorsal fin`, something which you would expect from a whale rather than a plesiosaur or pinniped. However if the animal was `splashing around` it could have been that the dorsal fin was really a displaced flipper. Even so, it leaves us no better off in deciding just what the creature really was.

• Head, Neck and Whiskers

Time to move on a bit now.

In 1943, off the north west coast of Florida in the Gulf of Mexico, Thomas Helm an ex–marlne, was sailing with his wife on their sixteen-foot yacht when at about 4 p.m. he had to change course for fear of hitting an odd animal that suddenly appeared and began making its way towards them.

The water was apparently as smooth as a mirror at the time and they were both able to get a very good view of the animal, which enabled them to provide the following detailed description.

> *" It had the head about the size of a basket ball, on a neck that reached nearly 4ft. out of the water. It was unmistakingly some kind of animal. The entire head and neck were covered with wet fur, which lay close to the body and glistened in the afternoon sunlight. When it was almost beside our boat, the head turned and looked squarely at us. My first thought was that we were*

seeing some kind of giant otter or seal, but I was immediately impressed by the fact that this was not the face of an otter or seal".

Helm had a keen interest in zoology and was apparently familiar with seals and otters.

"The head of the creature, with the exception that there was no evidence of ears, was that of a monstrous cat. The face was fur covered and flat and the eyes were set in the front of the head. The colour of the fur was uniformly a rich chocolate brown. The well-defined eyes were round and about the size of a silver dollar and were glistening black. There was evidence of a flattened nose, and just below where I judged the mouth should be, a moustache of stiff black hairs, with a downward curve on each side".

He further contemplates;

" Seals and Sea Lions have long pointed noses, and the eyes are situated on the side of the head like those of a squirrel or rat. The creature my wife and I saw had eyes which were positioned near the front of the face like those of a cat."

 To begin with, the animal is a mammal; it has fur and whiskers and is likened to an otter or seal. The head seen quite clearly does not appear to have ears, unless they are sleeked back. If they are not present then we can rule out an otter or otariid. A sea otter averages about 6 feet. in length, while a marine otter may manage 4ft., making the head and neck of the creature seen, practically the same overall size of the animal. These mustelids also reside in the Pacific, not the Atlantic. Although mistakes can be made, Helm presumably would have been able to identify an animal he was familiar with and despite sounding very seal like he makes much about the eyes of the creature being positioned differently to those of a seal. To me though, this description is appropriate for some pinnipeds such as phocid seals. Could Helm and his wife have seen a rare, or unusual seal, such as the then already rare Caribbean monk seal? The habitat of this seal ranged throughout the Gulf of Mexico from the Bahamas to Central America and it was known to be very approachable in nature, one of the things that led to its demise as sailors, whalers and fishermen culled it to extinction.

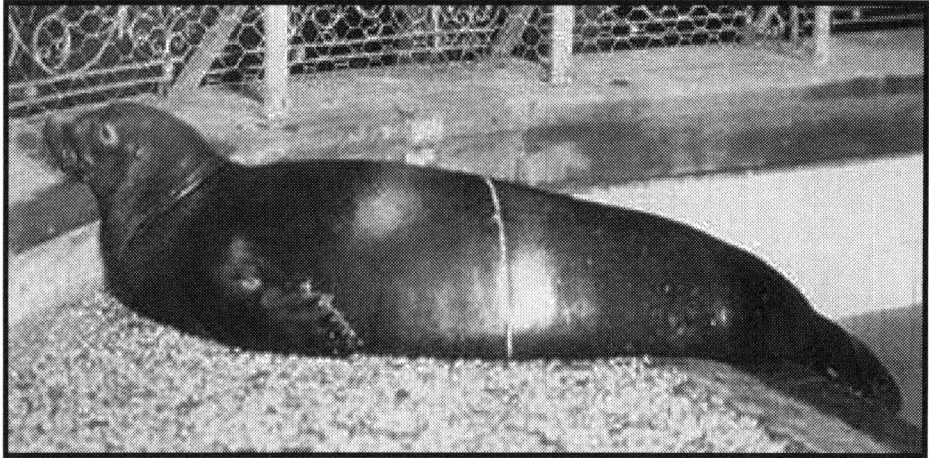

Fig.15 Caribbean Monk Seal

Although unconfirmed reports of survival have been made, recent surveys have concluded that this species no longer exists, being officially declared extinct in 1996. The last valid account of such a creature is accepted as being 1952, (*after* Helms` sighting) and because of its rarity it would have been unfamiliar. According to scientific description, it did have brown fur and reached 2.5 metres in length (8 ft.). Figure 15 in comparison shows the apparently similar looking Hawaiian monk seal, which could be viewed in its photographed stance as `cat like`. [15]

Despite all this there is however one fact that needs to be taken into account which may discourage such an answer.

Helm states that the creature is close to the boat and if the side of the boat was used as a reference to height, then we have a four foot head and neck, with no evidence of fore flippers, or disturbance of water, indicating such. This is quite a height and if a known pinniped were stretched out in such a manner, surely the rest of the body would be fairly noticeable.

Helm who later went on to write his own book on sea serpents recounted: *" It stared at the boat for an instant as if quite uninterested, then swung round and dived."*

A similar account was related to Heuvelmans by a friend of a friend, Michael Peer Groves*, who witnessed a strikingly similar animal in The Isle of Man in 1928. Peer Groves provided the following sketch:

---

* *Roland Peer Groves (son), verified his fathers account via my website. My memory is a bit hazy but I think that he mentioned that a photograph had been taken but had long since been lost.*

Fig.16 M.P.Groves sketch

The animal was described as having a head the size of `a large bull, but rather broader between the ears`, ending in a long, dog like snout.

The similarity to the Helm creature is amazing, but there are very few details apart from the very descriptive sketch provided. It is quite possible that the creature was just a common pinniped although again it has ears, which must make it a wandering, or non-native species. Peer Groves apparently commented:

" It was so nice....so nice. "

So nice.

The general feel of these two accounts may possibly remind the reader of the common demeanour and expected behaviour of a seal, shy, inoffensive and displaying the habit of mammalian curiosity.

For another head and neck encounter, again little detailed, but possibly showing a similar disposition we shall stay in English waters, but travel to Herm in The Channel Islands. It was related to Tim Dinsdale* by a Mrs H. Bromley and is taken from Heuvelmans book.

The sighting occurred in August 1923 and was witnessed by no less than 14 people (!), guests of Sir Percival and Lady Perry.

The guests in the company of a local seaman had decided to take a stroll along the beach, following the tide as it went out. They were equipped with hooked sticks in case they came across any lobsters and had been walking for some time they came to a large tide pool:

---

* *Dinsdale authored several cryptozoological books and was one of the greatest proponents of the plesiosaur theory. He wrote extensively on the subject of the Loch Ness Monster and even succeeded in filming `the monster`.*

> *"..but what held us all spellbound were marks on the sea*
> *weed as though something huge had come out of the pool*
> *and dragged itself over the seaweed covered sand, away to*
> *our right. We one and all turned and followed the drag*
> *marks, (if I can remember rightly), for some considerable*
> *distance, and then we came to a large pool, much larger*
> *than the first into which the drag marks disappeared. We all*
> *stood amazed, 14 of us, what could it be?*
>
> *Then slowly, away in the middle of the pool, a large head*
> *appeared and a huge neck, but we did not see the body;*
> *there it stayed with its great black eyes gazing at us without*
> *fear, then slowly sank back into the water."*

The group joined hands and tried to move into the pool in the hope of disturbing the animal further but their ever vigilant guide said that the tide was coming back in fast and that they would have to go back.

Fourteen witnesses! Did they all fail to recognise a known animal?

Further details given to Dinsdale, revealed a 3-4ft. neck and a wide mouth like a "sea elephants".

The animal had black looking skin and moved in an apparent slow and ponderous manner not seeming to be at all concerned with the attention that it was being given. We do not have any detail as to whether, ears, fur or whiskers were present but if Mrs Bromley compared the head to a seal then maybe they were.

At second glance there may at least be some similarities to a pinniped as Mrs Bromley thought the head of the creature was reminiscent of a seals, although she stated that she new what a seal looked like and said that it was not a seal. The mouth however did remind her of a `sea elephant` while Mr Toby Bromley who was also present felt that the whole creature looked like a gigantic eel, which is obviously quite a discrepancy..

Pinnipeds are well known to seek out and use tide pools, especially when they feel heat stressed (presumably the August day was sunny), and the 5-6 ft. drag marks are probably the sort of size marks that would be left by a large pinniped, (phocid?). Once again though, we have a 3-4ft. head and neck, above water, with no obvious sign of a body. Such a length would not be characteristic of a known pinniped or other sea going animal. Heuvelmans thought that the account provided a good description of one of his `long necks`, the tailless megophias.

The next report shares certain similarities to both the Peer Groves and the Helm account.

It took place in September 1957, as detailed by Heuvelmans. The sighting was made from a scallop ship, The Noreen`, as it sailed 120 miles east of St Georges Bank near Nantucket. The ships cook provided most of the information that was subsequently printed.

An animal had surfaced about 100 yards starboard of the ship with a `peculiar look about him`. The body was seen to be large with an alligator like head. The neck was of `medium size`, apparently matching the size of the head, while the body was shaped something like that of a seal with a mane of bristly hair or fur. This seemed to run down the middle of the head. The creature kept submerging then reappearing and the body was estimated at 40 ft. Despite the boat changing course the animal continued to display itself and when it surfaced it would turn its head towards the crew in a manner that seemed both `playful and curious`. During one of these precious moments two flippers, similar to those of a seal were also glimpsed. The sighting was reconfirmed twenty years later by the ships captain who could still not reconcile it to other animals that he had seen and commented that it seemed to stay on the surface gliding along with the ship.

According to Heuvelmans an estimate of twenty-six feet was given for the length of the neck!, making the neck nearly twice as long as the body, which seems quite absurd. Heuvelmans felt that the length from his original source was a typographic error and the length was really 2ft. 6 although conversely this does not seem that great.

The creature appears to have been very seal like in appearance and behaviour. Again, apart from the estimated size, the creature could be taken as being just such an animal. Although why experienced seamen the world over do not identify such characteristic morphology with such creatures is a mystery in itself.

So from this brief selective sojourn can we begin to form any conclusions?

Stoquelers report appears promising while the Tasmanian account appears to be some form of seal. The Murrumbridge animal is an oddity although it has fur and therefore must be a mammal.

The Scottish and Irish reports as well as those from Alaska and North America seem to fit loosely with some form of long necked seal. The Suffolk sea serpent if it was not horse or deer swimming must be some form of pinniped while the Helm and Peer Groves encounters clearly denote some form of seal. The Noreen account also seems to hint at such an animal apart from its size while the Herm description sounds seal like in the animal's attitude and demeanour. Unfortunately although drag marks were seen there is no mention of tracks, which may have indicated limb movement and could have then been helpful in identification.

Hopefully the reader may now appreciate why the title for this chapter was chosen.

Before we leave it in search of more detail I would like to include one last, again previously unpublished account that seems to sum things up quite well again proving very frustrating. It was emailed to me via my website by an American couple who I will refer to as the Ws. In their correspondence they made it clear that they were not publicity seekers, drug addicts or susceptible to hallucinations.[16]

> " My husband and I live in Northern California and just this last labour day weekend, visited some of our northern California beaches. Early on the morning of September 1, 2001, we were beachcombing an almost deserted beach (it was 7a.m.) and happened upon a creature, which my husband thought, was a seal. I had my doubts because of its colour, and its long and very limber and coordinated neck. Its face was also more pointed than most seals, it appeared to me to have an ermine-like or mink like face. I could not get close enough to it to really see if it had whiskers; they were not visible. The creature was sniffing the air at low tide, on the sand; it did not appear to be ill or anything of the sort. The colour was similar to that of a dark palomino horse; in other words, its fur was golden brown, rather than the usual dark brown, which seals have. It seemed also to be able to move much faster and in a more upright position than a seal. (We walked up a certain distance, where upon the creature would move a few feet away: we did these two or three times). The " flippers" did not appear to be exactly like those belonging to a seal. I knew that what we had seen was probably rare, since I had never heard of such an animal anywhere, either extinct or currently living"

Here we have a very seal like animal that seems to differ from known pinnipeds by virtue of its `long and limber neck`. It is witnessed by a non-attention seeking couple who are obviously familiar with seals of which there are plenty in California.

Incidentally, my website dealt *only* with the notion of a long necked seal and was never popular (!). This would seem to indicate that the couple were trying hard to reconcile their description with the nearest match that they could find and must have searched specifically for something that was close to what they had seen. Needless to say I hurriedly returned an email thanking them for their account as well as asking if they could possibly supply more details, such as size, the definite presence of fur and more exact movement details. Unfortunately I never received a reply.

The Ws., summed up their account and ended the email with a very apt sentence.

"Perhaps this beautiful creature can some day be identified".

I hope so.

4. LAKE CONFUSION?

While it is one thing to suppose that there are probably many unknown animals swimming the worlds seas and oceans it is quite another to believe that long necked creatures be they plesiosaurs or pinnipeds, have taken up residence in lakes the world over becoming the culprits behind countless reports of lake monsters. We will discuss some of the reasons for this view presently.

The following chapter has been included more as an afterthought, rather than a part of the original plan for this work. This is mainly due to the fact that lake monster accounts have been covered much more comprehensively elsewhere and include reports of apparent unknown animals no matter how unrevealing. This type of approach will not benefit this present work as to illustrate the case for a long necked pinniped we need to have more precise information at hand.

There is no getting away from the fact however that long necked, unknown animals very similar to those that have been reported at sea have also been variously reported in lakes from around the world. This means that we now have a new set of variables and questions to answer not only with regard to their possible identity but also as to why if these reports are genuine they favour such places. For the sake of completeness then this chapter will try and take a logical approach to some of these questions.

The included accounts have been chosen not because they clearly seem to identify a mammalian culprit but because they reveal interesting features rather than a description of just "humps", "bumps" or "upturned boats", which may have some relevance to the present work. Unfortunately this does mean using a selective filtering of reports, which in turn will reduce the amount of accounts that can be presented. This may appear quite one sided to the reader but I make no apology for this. There are many other books out there to read on the subject - see bibliography.

- ## Everyone's Favourite Monster

Many of the following accounts originate from Loch Ness, not so much because it sets a template for other lakes around the world but more because it allows ample opportunity to identify probable pitfalls with the whole idea of freshwater habitation. To avoid possible confusion I will mainly stick to a chronology of events and have followed Costello for most of the source material although I have used books by Witchell and Harris for further reference.

An early account that is generally believed to have taken place sometime in 1880 concerns an Eric Bright, who reported a sighting many years after it actually happened while he was on the Loch side near Drummadrochit. He apparently saw a large grey animal emerge from a wooded area waddle down a hillside on four short legs and then plunge into the water. He described it as having a long neck with a small head and it left a sizeable wake in the water as it swam off. Although he was only nine years old at the time, it obviously left a clear impression.

Sometime later, during the course of the First World War, a Mrs Cameron who was then a young girl, and her two brothers and young sister, saw a similar creature.

They were waiting for some friends, skimming stones on the water, when they heard a loud crackling in some trees on the other side of the little bay where they were playing. The crackling got nearer and nearer and they thought that something big must be moving. They were right. After a while a 20-foot long animal appeared moving like a "*caterpillar*". As the thing was face onto them they could not distinguish whether the neck was long or short but discerned a body that was the colour of an elephant with a shiny looking skin. Under the creature they saw two short, round feet and as it entered the water it lurched to one side and dipped one foot in the water followed by the other (how dainty).

Half a decade later in 1923 a chauffeur by the name of Alfred Cruickshank was motoring down the northern shore of the Loch. It was early morning and things were still a bit dark when he hit a bend in the road and his headlights picked out a large moving object ahead of him at a distance of about 50 yards. It had a large humped body about six feet high with its belly trailing on the ground and was about 12 feet long. A tail was present that appeared to match the length of the body making it something like 20 feet overall. It again waddled across the road using two visible legs and as it did so he made out the head of the creature, which was set right on the body with no apparent neck. The head gave the appearance of being "pug nosed" and he thought the overall colour was dark olive to khaki.

Although all convenient preludes it would still be some ten years before `Nessie` would explode onto the scene and go on to shape much future thinking and assumption with regard to lake monsters. Things of course really took off in the early thirties and 1933-4 were very good years to stake a claim to notoriety.

In 1933, the MacLennan family witnessed a strange creature in the loch. They had caught glimpses of the monster on a couple of occasions previously that same year. On this particular occasion Tom MacLennan was able to describe a 30-foot long animal, which gave the impression of having four flippers while his wife was able to view the front and rear end of the creature. It had a head that was held about three feet out of the water on a tapering neck, the head being little wider than this. On the back of the neck "*hair*" or "*wool*" was seen. The tail was apparently divided like a fish, which is slightly unusual, but Costello took this to mean that it was divided in two and hence was really the rear flippers of the creature rather than a *bona fide* tail. This appears to make good sense as otherwise we seem to have a new type of animal, a long necked fish which does not seem to have been described elsewhere. It is not clear how much "wool" or "hair" was present but this feature instantly brings to mind the mane sometimes reported in some sea serpent accounts.

So if we take on board Costello's assertion of the tail, we have an animal that seems to match its sea going cousins. It is big, has a long neck and possibly a mane. Mrs MacLennan later went on to have a further strange encounter with a slightly different creature in August while she and her family were on a return journey across the Loch from church. She had wandered down to their boat on the shore of the Loch when she came upon a grey looking animal resting on a ledge about six feet above her. After shouting for her family the creature lurched and slithered off the ledge into the water. Her family therefore did not get to see the occurrence only some after disturbance in the water afterwards. She later described the beast as being hunched up, rear facing her with its head thrown back. Apparently 20-25 feet long it had short, thick clumsy legs with a kind of hoof like a pigs. This description was later revised and subsequently described as being more like a dinosaurs feet, cloven (?). It had lurched itself up on the two forelegs while the back legs stayed on the ground `seal wise`. There were no ears apparent and the back was ridged like an elephants. In her opinion the rear legs apparently did not look as if they could support the animal because it slithered along the ledge and into the water like a seal. She did comment however that it must have climbed "*like a monkey*" to get to its position on the ledge which does not quite tally with the obvious clumsiness that she attributes to it. There also appears to be no mention of a familiar long neck. A Mrs Reid, the wife of a local postmaster in December 1933, saw something smaller as she was motoring into Inverness. She caught sight of the animal about 100 yards distant, partly obscured by bracken. She was not definite about her description but thought that it had a thick hairy mane on its neck and generally seemed hairy. It was only 6-7 feet long and shaped like a hippopotamus, appearing to be a slow moving type of creature. It had a rounded head, short thick legs and was a very dark colour.

The next account concerns a very detailed and perhaps most revealing land sighting, which may give us more insight into what, we may be dealing with. According to Costello, the report was to some degree neglected due to the commotion that was being caused in the early thirties by the whole Loch Ness affair and occurred in January 1934.

Arthur Grant, a veterinary student was motorcycling from Inverness to Drummadrochit at about 1am. It was an overcast night but at the time of his encounter the moonlight had lit up the road in front of him. As he was riding he saw a dark object in some bushes to his right. As he approached this particular point in the road, something big bounded out in front of him. It managed to cross the road in two further bounds before vanishing into some bushes on the opposite side of the road by the loch shore. He later recounted that in those few seconds of confusion he got a pretty good look at whatever had appeared in front of him and went on to describe a long necked animal with oval shaped eyes on a small head which also possessed a 5-6 foot long, powerful looking tail. He estimated the total length of the creature at being approximately 15-20 feet, with his initial thought being that it was some sort of cross between a pinniped and a plesiosaur. He dismounted his motorcycle and proceeded to chase after the startling apparition making a note of where it had entered the loch with a great splash.

He was later able to give a more detailed description of what he had seen and described an animal with a head like a snake or eel that was flat on top, a large oval eye, longish neck and somewhat longer tail with the body being much thicker at the tail end. It was black or brown in colour and the head was about six feet off the ground at the end of a 3-4 foot neck. The tail he reckoned was 5-6 feet long and the overall size approached 20 feet. Luckily a party of students who happened to be on vacation at the Loch, who had also had their own sighting of two humps during their stay, were later able to acquaint themselves with Grant and spend some time with him at the location of his sighting. As a result they managed to ascertain some interesting further information.

After listening to Grant they managed to obtain a more exact description of the animals movement on land from the time that he saw it to its disappearance. It had apparently *"loped"* across the road using all four of its flippers, first putting down the front ones then arching its back and heaving the hind ones forward in the manner of a sea lion, the stomach not touching the ground. On closer examination of the sighting area, corresponding tracks were found that resembled "scrapes" or *"skids"* about 5 feet apart. One of the students a fellow in zoology later examined a collection of animals at the Royal Scottish Museum on his return to Edinburgh. He found that the track dimensions exactly fitted those that a fully-grown bull walrus would have left in similar circumstances. Grant had also managed to sketch the beast soon after his encounter and it was later redrawn showing a typical plesiosaur for Dr Ouedemans who was visiting the area hopeful that is theory would finally be justified. In Grants original sketch, which can be found in Costello's book, such certainty was lacking and he appears to have been undecided with regard to the tail of the animal, not actually including one in his drawing. Costello again felt that the tail was not really a tail but the hind flippers, which had initially been stretched out by the creature before the encounter. These then became obscured in the commotion that followed.

Again a reasonable assumption, but in hindsight, for an animal that moves in such a precise way it is hard to understand how it would have managed to do this with such a long tail unless it kept it pointing upright during its movement like an otter. On the face of it, if the animal had not had a long neck, a wandering tuskless walrus would seem to fit the description quite well. The walrus does move on land in the same fashion as a sea lion and does appear to travel widely. And one could not ask for a better description of the type of movement that it would make from Grants account. But *there* is that long neck.

Unfortunately this account may be somewhat tainted. Paul Harrison in his comprehensive guide to the Loch Ness mystery informs us that Grants report may have been part of a hoax portrayed in part by Grant in collusion with a self proclaimed big game and monster hunter one Marmaduke Wetherell. Wetherell who obtained backing from a national newspaper to hunt Nessie went about doing everything possible to show that she did in fact exist. He was involved in various publicity stunts and hoaxes and conveniently managed to find supporting evidence practically everywhere he looked. According to Harrison he did in fact accompany Grant the day after his encounter to the spot where it took place and began citing flattened shrubbery, footprints and even a dead goat as corroborating evidence.

However in his book, Witchell mentions that Grant initially visited the scene with his family, being joined only later by Wetherell but it is this possible liaison, which has brought the account into question. Grant of course was a veterinary student and must have been familiar with seals and sea lions and the fact that the track marks found conveniently match a bull walrus * and the precise method of reported movement on land could possibly indicate a hoax. On the other hand although the later drawing of his beast took on the familiarity of a plesiosaur it did not appear definite to begin with. He would have presumably been familiar with these too and could have made the account much more `classic` with regard to this in its nature, unless of course he had taken a fancy to a long necked seal idea via Ouedemans or Goulds books. Witchell† goes on to state that as a result of the sighting and the ensuing ridicule Grant was exposed to, he was forced to miss a term at college. Probably not the best of starts to a potential carer as a vet.

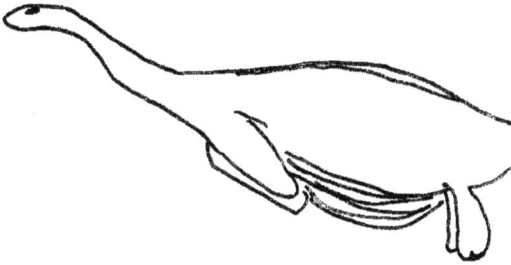

Fig.17 Grant's creature based on his original sketch.

Otherwise, away from the controversy, here is a report of an animal that looks a bit like a seal, moves like one and also meets the dimensions of at least one of its order. It also appears quite used to moving about on land, which is something that we will discuss in relation to lake monsters in more detail presently.

In February of the same year Jean MacDonald and Patricia Harvey, both girls at the time, described how under a full moon while walking along a road close to the town of Inchnacardoch they saw the fabled animal cross the road in front of them. They both described it as being 8-10 feet long with four feet and a head, which was about 6 feet

---

* It would probably not be a mature male elephant seal as there is no evidence of a proboscis and we would have surely have heard about `horned` dinosaurs if this had been the case

† The Loch Ness Story, Nicholas Witchell and The Encyclopaedia of the Loch Ness Monster, Paul Harrison (see bibliography).

off the ground. The body was thickest at the shoulder and tapered towards a tail. It was very dark in colour but white under the neck. The legs were very short and it moved rapidly and quietly in the direction of the Loch. A few months later in June Margaret Munro, a housemaid at Kilchumein Lodge watched an animal rolling on a shingle beach for about 25 minutes. Much of the body was clear of the water and it had a giraffe-like neck and absurdly small head for its body. Its colour was dark grey although again the chest was white. The skin was like an elephants and she saw two short forelegs or flippers. As she continued to watch, the animal kept turning itself in the sunshine and was seen to be able to arch its back into large humps. A sketch that she later made shows something similar to Grants although the hindquarters trail off into the water. This was then followed in July by an encounter made by Ian Matheson, who watched for about an hour as a beast emerged from the shallows of the Loch looking like a horse with five humps and with `the air of flippers`. He likened its movement to a worm wriggling. The five humps increased to twelve as the animal wriggled its way ashore and the head was noted to be smaller and thicker than a horses with the neck being "*heavily maned*". The animal, which appeared to be feeding on some plants in the area by the shore was 30 feet long, not very thick and actually went on to shake the water from its mane as perhaps a terrestrial horse would do.

Much more recently in 1979 Donald MacKinnon claimed a sighting of a strange creature that appeared from a wooded area then proceeded to walk down to the Loch, slithering into the water. It was grey, about 24 feet long with four feet that had `three fingers` on each foot. Despite the gap in years, Nessie was still coming ashore for some reason.

Worth making a note of from these accounts are the very visible limbs, something that I shall comment on later.

The information gathered so far seems to describe two different types of animal. One appears to have a long neck while the other lacks it altogether. The movement on land also appears to show some discrepancy and at face value appears quite confusing. Or is it?

Some of the accounts that describe a dragging movement also include a tail, which could in fact be trailing hind limbs, very reminiscent of perhaps a phocid seal. Several of the accounts are vague about a long neck and the overall size of the animals in these accounts appears to be on the small side. So are these just accounts of a large wandering phocid? We have already learned that pinnipeds travel widely from their usual habitats and a possible candidate could be a juvenile bull or female elephant seal, which had possibly strayed many miles from its usual habitat (which as we have seen they do). Elephant seals are the largest of the pinnipeds and would match the sizes reported (see footnote). Then there is also the possibility of a tuskless walrus again a large animal, or maybe some other species of seal. If the reports of manes are to be believed though then it would have to be an alien one.

As we shall see later there is at least some evidence for the possibility of seals travelling in this way sometimes for a specific purpose.

- ## Caught on the Hop?

A similar description of an apparent lake monster seen moving on land much like Grants, can be recounted next, this time originating in Ireland. It came to light after what appears to have been a photographic hoax, apparently portrayed in 1968.

A journalist was tasked with digging for some background information after a supposed picture of a dinosaur type animal had been taken at Glengarry or Sraheens Lough, which is situated on Achill Island off the coast of Co. Mayo., (incidentally Achill Island was renown in the past for various attractions including seal caves).

There had in fact been local tales of an unknown beast inhabiting the vicinity for some time before the incident took place. In the course of the reporters research he happened to interview a local 15 year old lad by the name of Gay Denver, who described to him an unusual encounter he had witnessed some 3 weeks before the picture was supposedly taken.

Mr Denver had been cycling home from mass and had dismounted his bicycle by the lake when he saw an animal about 50 yards away from the Lough which seemed to be *"humping"* itself along and climbing a nearby turf bank. It was bigger than a horse with a long and slender head like a sheep. It moved in a jumpy way like a kangaroo and had a long neck and tail. The hind legs were bigger than the front ones and it was about 12 feet long.

The `jumpy` way the animal moved is interesting, as it seems to indicate movement that may be close to Grants description and seems to indicate structured travel, especially if it were climbing. From the seal point of view this may have presented many problems for a phocid, but an otariid would find the going a lot easier and depending on how steep the bank was such an animal would need to lift its hind flippers after placing its fore runs perhaps in some motion that may appear to be kangaroo like. Denver does however report a tail, if that's what it really was, but like Grants it seems hard to reconcile it with the movement described. If however rumours had been doing the rounds about some Irish `dinosaur` being seen locally, this may have influenced his description.

Like Costello who was not convinced about tails, I have to agree that such descriptions of such a distinguishing feature may occur simply due to the fact that the witness thinks they are seeing a plesiosaur, which most people are familiar with in shape.

Further south along the coast from Achill, lies Connemara. This area has long been linked with the fabled Irish `horse eels`, creatures that were described in the nineteenth century as resembling conger eels but with a head like a horse and matching mane. Rumours and sightings of such creatures have been made well into the twentieth century and although they do not quite match typical lake monsters in size and general appearance, being more otter or eel like, the reports come from an area that comprises many small pool like lakes which form a chain to the coast. Several expeditions to this area have been made in search of the elusive `horse eels` but have unfortunately not revealed any further clues to their possible identity.

There have also been persistent rumours of a semi mythical animal known as the dobha-chu, which is believed to be some form of large otter or `master otter` and it could be possible that such a creature may be responsible for many `horse eel` reports. Otters are notoriously elusive and a form of giant otter, such as those found in South America may well be a `monster`. In an excellent article discussing this possibility (see below), Gary Cunningham cites some convincing incidental evidence involving the possible survival of a prehistoric otter known as *Potamotherium,* albeit in a modern form.[1] A similar animal has also been rumoured to reside in Scotland and certain aspects of an otters appearance would correlate quite well with some of the morphology attributed to the Loch Ness creatures (`white stripe`, long tail). *Potamotherium* was also more adapted to a life in the water than present day otters and would present another alternative to the plesiosaur identity for lake monsters. Unfortunately although a larger species could have evolved, *Potamotherium* was only about 5 feet in length and presumably lacked a mane.

It is also probably also worth remembering Howard St. Georges sighting off the coast of something more plesiosaur like.

I will revisit the subject of `horse eels` later with regard to the pinniped side of things but before we move on I will briefly include another unusual sighting of an animal seen in Connemara this time from Lough Skanaveer. *

It occurred on a misty and rainy morning in 1944. A Mr Canning was walking down to the lake to fetch a pregnant mare. As he got near to where it was, the mare stood up and Mr Canning caught a glimpse of some animal circling the mare from behind. Canning initially thought that this was the foal, which had therefore already been born. As the `foal` was near to a small stream that flowed from the lake he was worried that it may fall in and drown so he began to approach it to ensure that this did not happen. As he got nearer however, the beast, which he thought was the foal, apparently detected his movement and promptly disappeared into the water. On reaching the scene he found that the horse had not in fact given birth (although a few days later she did), and after realising his mistake tried to reflect on what he had witnessed. When later interviewed about the encounter he stated that there had been some unusual things that he had noted

---

* Richard Muirhead in an article for Animals and Men, issue 14, came across an interesting article concerning and animal that was caught in a trap in Slane, County Meath in 1869. It was the size of a good cat covered with long wiry hair. It had a sharp pointed snout like a weasels and a mouth that showed four "large tusks", two facing up and two down. It had a small mane of dark brown hair down the whole length of its back and twelve toes or claws on each foot, seven on the outside and five on the inside. It was stouter than a cat and was dark brown with white on its breast. Was this a Dobhar -Chu?2

about the foal like animal. In his words *" It was long...rather a bit high. It was black. The neck seemed a bit long... "* He later claimed that the animal had ears and legs and circled the horse "gently". When subsequently asked about the presence of a tail he stated that he had not seen one. He was convinced that it had not been an otter.

• Swedish Odyssey

Jumping next (or should that be loping), to Sweden now, Costello also included some lake monster reports from Lake Storsjo, the deepest lake in Scandinavia, situated in central Sweden. 0This appears to have been a good place for a monster as it even had its own little island with a runic stone supposed to protect the local inhabitants from a terrifying serpent which was depicted on it.

The mystery beast that inhabited this lake appears to have been the Swedish equivalent to Nessie in later years but the two reports that I will include both occur in the nineteenth century and should therefore be a bit more unbiased as to possible identity. Unfortunately though in the first account things will get slightly weirder.

In 1893 two girls who were both described locally as truthful, were washing their clothes in the lake when they saw a large animal swimming in the water, which proceeded to stop some distance in front of them. The head, which was large and round, rose and fell in the water for about half an hour while they watched. Short fins larger at the rear than the front could be seen and the animal had large sized eyes and two clipped looking things on the back of the neck. Its colour was grey with black spots. One of the girls decided to throw some stones at it, which made it start towards them and it was now that the girls, bless them, decided to flee. One ran to a nearby railway line while the other climbed up a tree.

Despite their obvious fright they were later able to give more details.

The head of the animal was round and like a dog with eyes as large as saucers. The mouth was open and a tongue could be seen flicking about inside. The eyes were 6-7cm apart (presumably converted to decimal form from the original report) and the head was 3 feet long and wide. The neck was 8-9 feet long and the back about fourteen. On the head were two big ears, which were laid back along the neck and there were two objects on the neck that could be extended and when the animal swam away its fore feet stroked the water.

Although we have the usual basic similarities to some of the other reports this one now adds some extending appendages on the neck, something new and rather strange.

Although some other sea serpent and lake monster reports allude to `horn like protuberances`, rarely snail like stalks and even neck appendages referred to as "frills" or "kippered herrings", apart from Heuvelmans notion of snorkels, what the girls appears to be a unique attribute. However as there does not seem to be any creature aquatic or terrestrial that comes close to the description and the fact that the girls watched the thing for half an hour, a good length of time for observation, it would appear that there may still be a few surprises in store when dealing with apparently familiar lake monsters. *

As a result of the encounter a Norwegian adventurer was brought in to bring the monster to heel, although all efforts proved fruitless.

Later in 1898 a newspaper researcher and five other people witnessed an animal about a kilometre from a house on the shore. It swam against the wind and turned allowing them to glimpse the belly and two fins. The body was about four feet high and about one and a half feet wide. The length may have been 4-5 metres the head and neck being one and a half, although the head was not raised fully out of the water.

Mention was made of ears, which were white, and two feet apart above the water. The body was generally smooth but had warts on it in places being a sort of cinnamon colour, apparently slimy or scaly. There were green things hanging off its neck, which were described as either being a mane or waterweed. It quietly moved off and they observed it for a further half an hour before a steamer somewhere in the lake blew its horn at which the animal cocked up its ears appearing frightened and sank out of sight.

Now a kilometre seems a long way for witnesses to describe such detail which they also seem to contradict in their observations (slimy, scaly, smooth), but of interest to us are the ears which definitely seem to be ears as they react to sound. In fact a similar animal was reported on two further occasions both of which also described cocked ears.

Moose are common around many Swedish lakes and immature specimens, `in velvet`, may not be readily identified for what they really are. Speculation that the Storsjo beast and similar sightings are in fact just these animals swimming does hold some weight. But although the girls in the 1893 encounter probably made much more of their experience, they describe an 8-9 foot neck which appears a bit excessive for a Moose.

Fifteen years before the girls had their sighting, a local sawmill mechanic, Martin Olsson, witnessed something similar.[3] He was fishing near the island lake, Forso, when he got the feeling that he was being watched. He looked behind him and saw a 6-foot snake like neck with a large disproportionate head on it. A hairy fringe hung down the animals back apparently stuck close to the neck due to wetness. It was a blackish rust colour with skin resembling that of a fish and had two humps in the water behind it. Moving cautiously Olsson decided to paddle away from the creature but he only got about 10 metres before it began to swim after him. He stopped rowing and the beast sat in the water staring at him for about five minutes. The strange creature eventually

---

* *The exception with regard to the theory of exploitation in the cited examples however would be Bear Lake. Although it is connected to a river, the river actually flows into another lake The Great Salt Lake, which apart from being unable to support anything more than shrimp does not connect with the sea. As both lakes are in the heart of Utah any monsters would probably be better off trying to fly to the sea than struggling overland.*

dropped beneath the water allowing Olsson to breathe a sigh of relief.

An interesting but very vague land report comes from Sweden's neighbour Norway, in again 1893. It describes a strange beast seen for the last time in Fjoesvika. It took the form of a 15 metre black animal with long head and mane on its back. It was apparently witnessed moving from a forest and plunging into a nearby stretch of water. [4]

• Not to be outdone...

Europe and Scandinavia are of course not the only places to spy lake monsters.

North America over the years has become a veritable hot bed of such activity.

There are many lakes on the continent, which are reputedly inhabited by `monsters`, but in general descriptions of them tend to be less detailed than their English cousins. Again there are long standing elements of cultural superstition that tend to confuse things.

I will however include two reports from North America to represent their case but again the reader is recommended to view the bibliography section for further research. Both of the accounts originate from Bear Lake in Utah and I will not dwell on them too long as detail is sketchy and their could be other culprits ultimately behind them.

The first comes from 1868 when a head resembling that of a serpent was seen in the lake by some locals. It was covered with light brown fur like an otter and it seemed to have two observable flippers, which were compared to fishermen's oars. Then in 1874 four fishermen saw what they thought at first was a large duck on the lake. When it got nearer however the face was seen to be covered with fur or short hair of *"a light snuff colour"*.

The face was apparently flat, very wide between the eyes, which were very full and it had prominent ears like those of a horse but not as long. The whole head reminded them of a fox with the distance between the eyes being similar to that of a common cow. It apparently did not look ferocious and was not in any great hurry to leave the sight of the encounter.

A duck, a horse, a fox and a cow, a rather unusual combination by anyone's standards.

• Galloping Monsters?

Lastly it is worth including a report from Iceland, again a place with a rich folkloric history.

In 1984 two bird hunters on the shores of Lake Kleifervatn observed two amphibious monsters emerge from its waters. They were apparently black and quite horse like in their appearance although they were bigger than normal horses. After emerging they proceeded to cavort and frolic on the shore for an undisclosed period of time, moving like dogs, before returning once more to the depths of the lake.

This report, maybe for good reason, has been brought into question due to the fact that some time after the encounter footprints taken to be those of the animals were supposedly found revealing cloven hooves. *

Anyway if the account is true then we have two unusual and apparently merry creatures capable of gambolling about on land quite easily.

- **Pushing the Point?**

Obviously there are many more accounts that could be included here but as the reader may appreciate things are from being clear cut when it comes to evidence for lake monsters. Perusal of more comprehensive information on this subject will probably help their case but hopefully the reader will see the significant similarities that can be made. Even from the small number of accounts that I have included some of these animals appear to look the same and share the same sort of physical characteristics and morphology, (an exception may be a monster from Lake Mendota known affectionately as `Bohzo`, who apparently gave a young girl a bit of a surprise in 1917 when it began to lick her foot underwater, before surfacing with a dragon like head and *"friendly, humorous look in its eyes".*)

So the real question that we need to ask here then is why do very similar animals appear to inhabit different aquatic environments, are they the same species or are they different ones?

First up although many of these lakes are linked to the sea by rivers etc. they are enclosed environments so unless the animals behind these reports are adept at travelling over land, sometimes quite vast distances to get to them, or are smaller than commonly perceived allowing them to swim up connecting rivers and lakes undetected, then there must be breeding populations that exist or have existed in these bodies of water. This in itself poses a bit of a problem for both the plesiosaur and pinniped theories as no matter how much evolution may have potentially changed them, both animals would need to surface at regular intervals to breath thus surely making them more obvious in the process. If breeding populations do or did exist to account for their continued lineage over the years a number of animals must be present, again making long term concealment difficult to accept.

---

[4] *Cloven hooves and tracks that look like such have been linked with some descriptions and sightings of mysterious water creatures and are a bit of a mystery in themselves if the animals have evolved an aquatic existence.*

This rather thorny issue can possibly be countered by stating that the lakes in question are big places and that the creatures giving rise to the reports may simply periscope their heads and necks undetectably above the surface every now and then to take air. Also perhaps that evolution has allowed the development of a more advanced aquatic existence, reducing their need to rely on land or the surface of the water, which in turn helps them keep hidden.

However if for instance we take at face value the reports of such creatures seen on land, such as those at Loch Ness then such animals seem to need to come ashore for some purpose. Again this should make a breeding population much more conspicuous and would seem to indicate that they cannot have adapted that completely to an aquatic environment in the first place, QED.

In fact a few more reports from Loch Ness clearly seem to demonstrate this need for land excursion and seem to indicate that the creatures are more than capable of accomplishing it.

In 1933 a William MacGruer related to the *Inverness Courier* how five or six children under the age of ten had gone to hunt for bird's nests along the shore, circa 1912-1919. They had not gone far when they saw a queer looking creature emerge from some bushes and make for the loch. It reminded them of a camel although it was shorter but of the same colour, a sort of sandy yellow. It had a long neck, humped back and fairly long *legs*. Then in 1936 a Lieutenant-Colonel Guy Liddell wrote to the *Times* describing an account of the beast seen on shore related to him by a Mrs Peter Cameron, the warden of a local youth hostel. In 1919 when she was fifteen, she and her two young brothers on a sunny September day, saw an animal on a marshy shore " loping" its shoulders and twisting its head from side to side. It too had a long neck, small head and four *limbs*, which along with its colour also reminded her of a camel.

Were there any escaped camels slumming it at Loch Ness between 1912-1919?

In both these cases and most of the other Loch ness reports that I have included, definite limbs are reported which appear to be able to propel the animals in question, in some instances, fairly efficiently on land. This would seem to indicate that they obviously need to preserve this function for some reason. In fact if the Storsjo accounts are accurate which describe ears, then again it would seem to confirm the notion that they have not evolved too much for aquatic life as advanced marine mammals such as the cetaceans and phocids have lost their visible external ears making them more streamlined for swimming.

Why then if such amphibious creatures are truly confined to lakes have they not been caught?

Secondly there is the need to obtain sufficient quantities of food to support a breeding population if it is present. Loch Ness, arguably the most persistent and popular place for a long-necked leviathan, is actually quite barren on the food front, as are many similar reputedly monster-haunted lakes. Research in the early 1990s proposed that if a population of ten animals, enough to form a breeding colony, were living in Loch Ness then due to the relatively small volume of fish attainable, each animal would not be greater in size than 300 pounds, hardly the sort of size for a monster whether seal or plesiosaur.[5]

In fact, there may only be certain times in the year, such as fish spawning seasons, when enough food to sustain the necessary populations would be available and unless the creatures were vegetarian the logistics for their continued existence just do not seem to add up.

The reader may argue that the sheer volume of reports from Loch Ness and other areas indicates that there are thriving populations, but before jumping to such conclusions we have to remember that not every hump cited as a monster may necessarily be one and we need to take into consideration the question of mistaken identities once more. This is no easier when dealing with isolated bodies of water.

With the exception of modern whales, sunfish and oarfish, which we can probably exclude, we still have known seals, deer, moose, elk, horses, sheep and cows and now otters. There are also very prehistoric looking sturgeon and eels to contend with both of which can attain large sizes and in the case of eels can actually move on land. If you combine these with the mythic, beautiful landscape surrounding such places the cultural mythology and perception of what a monster should be, numerous boat wakes, vegetable mats driftwood, seismic disturbance etc. we are surely asking for all sorts of descriptive trouble.

- ## Exploited

An alternative solution to the premise of breeding populations, which seems slightly more reasonable to me, is that of exploitation. In this scenario such creatures do or have at least in the past, travelled inland to exploit food resources that are available at certain times. They are or were capable of traversing long distances by sea and land for this purpose. To me this makes more sense and here the long necked pinniped theory holds more weight, for today as they have done in the past, seals *do* exploit food sources in a similar way and *can* move overland for considerable distances. The late Dr Gordon R. Williamson, documented seal sightings at Loch Ness and in other Scottish Lochs.[6] He quoted several instances of seals being seen in Loch Ness; one in 1933, two in 1934, four between 1972-1980 as well as an adult and 4 month old harbour seal in 1985. In fact for a seven-month period in 1984/5 a harbour seal (the same one?), took up residence in Loch Ness, the first documented incident of this nature. Whether it entered via the Caledonian Canal or overland is not clear. There had also been a sighting in 1895 from Loch Oich while further research also showed that Loch Shiel, Loch Hope and Loch Maree had also been frequented by seals in the 1980s, (five in L. Sheil). Although a hydroelectric Dam now blocks the river outlet in Loch Awe Williamson also included a brief quote from 1793, which told of how *"The seal comes up from the ocean into Loch Awe in quest for salmon"* and how in ten years up to 1883 seals were seen in the loch almost every year presumably doing just this. More recently in 2002 "Andre" an errant seal was actually awarded a fishing permit after he became trapped in the river Leven near Loch Lomond and began to take advantage of the salmon stocks available there.[7]

This means that seals were and are able to travel frequently into Highland Lochs and therefore some of these instances could quite easily explain reports of highland lake monsters. This also obviously ties in quite nicely with the possibility of inland exploitation. Seals are well known to take advantage of such food supplies. In fact "jail bars" which are 36ft tall and 12 feet wide have been erected at the entrances of fish passage structures in Bonneville in Oregon, USA recently. [8]

The purpose of these is to prevent sea lions from devouring salmon. Rubber bullets and underwater fireworks had also been used as deterrents, but the sea lions have still managed to chow their way through 2,500 salmon, 4% of the total number of salmon that have reached the dam so far. One clever individual known as C-404 actually managed to push its 20 inch wide body through a 15 inch gap all the way to a fish counting window despite being shot three times with rubber bullets. In fact possibly as a result of this deterrent action sea lions have now been spotted many miles up a nearby river looking for food[9].

Another seal, nicknamed Sammy, has taken up residence in a North Yorkshire river some 60 miles from the sea apparently taking advantage of the rich fish supply.

So what happened at Loch Ness in 1933/4? Could some presently known pinnipeds have been using the Loch as a convenient stop off point? Did a known or possibly alien pinniped take residence in the Loch during the First World War or the early thirties? Or did an altogether different type of pinniped find such a journey beneficial?

Loch Ness is of course connected to other Lochs by the Caledonian Canal which also feeds into the sea and both Lake Storsjo and Lough Glengarry are both in areas which have tributaries at least providing some sea going entrance or exit.

• Conclusions?

If we take Loch Ness, wrongly or rightly as an example for Lake Monster existence and the above reports from various places at face value, we can see similar challenges and discrepancies facing other such creatures in lakes the world over.

Obviously many of these lakes are much vaster than Loch Ness and the concealment of large, unknown animals easier and there are probably other animals responsible in some cases both known and unknown. However in smaller lakes, unless we are dealing with some form eel or fish that has grown a long neck, evolving into an amphibian, making it glimpsed at the surface only rarely, there are significant problems and flaws associated with the long term presence of plesiosaurs and long necked pinnipeds in them. Unless of course they are just passing through.

If such creatures were but travellers to these places for whatever reason, then there is more scope for their continued elusiveness and why possibly today with so many man made hindrances, reports are less in number.

Such exploitation may have also offered further benefits for such animals and I will come back to this hypothesis later on in more detail and more explicitly, in relation to the Surreal Seal question.

5. DINOSAURS THAT LOOK LIKE SEALS, OR SEALS OR SEALS THAT LOOK LIKE DINOSAURS?

" Returning from a hunting trip, Orde-Lees, travelling on skis across the rotting surface of the ice, had just about reached camp when an evil, knob like head burst out of the water just in front of him. He turned and fled, pushing as hard as he could with the ski poles and shouting to Wild to bring his rifle. The animal sprang out of the water and came after him, bounding across the ice with the peculiar rocking horse gait of a seal on land. The beast looked like a small dinosaur, with a long serpentine neck. After a half dozen leaps, the sea leopard had almost caught up with Orde-Lees when it unaccountably wheeled and plunged again into the water. By then Orde-Lees had nearly reached the opposite side of the floe; he was about to cross when the sea leopards head exploded out of the water directly ahead of him. The animal had tracked his shadow across the ice. It made a savage lunge for Orde-Lees with its mouth open, revealing an enormous array of saw like teeth . Orde-Lees` shouts for help rose to screams and he turned and raced away from his attacker. The animal leaped out of the water again in pursuit just as Wild arrived with his rifle. The sea leopard spotted Wild and turned to attack him. Wild dropped to one knee and fired again and again at the onrushing beast. It was less than 30ft. away when it finally dropped."

The above account, gripping in the extreme, is not an encounter with some unknown animal, but a chance encounter with a leopard seal, as related in Ernest Shackleton`s *"Endurance"*, a record of his Antarctic expedition. A similarly harrowing encounter with such creatures is related in 1980, when two Antarctic scientists, scuba diving, were confronted with three leopard seals that became unduly aggressive towards them. The divers had to try and defend themselves with two metre lengths of angle iron. The seals apparently repeatedly dived at these, coiling up their necks to strike at them as they attacked.

More recently in 2003, a 28-year-old marine biologist who was snorkelling near an Antarctic research station was attacked and pulled underwater by a leopard seal. [1] Despite her colleagues rescuing her, and attempting resuscitation she died as a result of this tragic and unusual incident.

The reason for including these short diversions is to consider another possibly related but altogether more fearsome long necked animal.

Roy Mackal in his book, *"Searching for Hidden Animals"*, related some very interesting information about an animal that is apparently the scourge of the native Eskimos around King Island in Alaska. The island lies 110 miles south of the Arctic Circle and is only two and a half square miles in area with a small Eskimo population. It is situated in the Bering Sea and is part of the Aleutian Island chain. [2]

The information was given to Mackal by a naturalist, Dr John White, who had apparently met and talked with the native people of this area at great length, obtaining a description of a frightening creature, referred to as Tizheruk locally and as Palraiyuk further south around the island of Nunivak. This fearsome animal was usually noticeable sporting a 7-8ft. neck and snake like head, which it would show out of the water and had a tail that had a flipper on the end of it. It was often encountered in bay areas and had been known to attack man on occasion. The natives could apparently detect its presence by ear and it could be called by tapping on the side of a boat, which as Mackal discovered, was a sure fire way of attracting a leopard seal, (should anyone wish to do so). Based on this information, Mackal speculated that a northern counterpart, to the Antarctic leopard seal might exist. If it did, Mackal thought that it might have evolved a more elongated neck while possibly losing its fore flippers heightening its reptilian appearance.

As already mentioned, The leopard seal, which already looks quite reptilian, fits the aggressive nature of such an animal, (although scientists feel that this behaviour is more often than not based on mistaken identity on the part of the seal), and Mackal concluded that the isolated nature of the locality and Alaska in general, could conceivably conceal such a creature quite easily.

Unfortunately, apart from the information given to Mackal by Dr White, subsequent detail from any source appears to be lacking. Indeed what further information can be gathered becomes rather confusing.

The Indian lore of North America and the British Columbian coast portrays many sea animal counterparts to the more familiar land ones, universal among them is one very similar animal to Mackals. [3] Generally seen as some type of `sea wolf' and variously named as Sisiutl, Wasgo or Haetelik / Hiyit`lik this belief seems to extend further north to Alaska where a similar beast is called either Tirichik (?Tizheruk), Mauraa, Nikaseenitluloyee or Palraiyuk. Petroglyphs created by the native tribes in these regions all seem to depict a similar animal. The Manhousat people of Flores Island and Sydney Inlet around West Vancouver Island apparently described Hiyit`lik as being a creature 7-8feet long and fast moving on land or in the water. It had legs but used its body rather than its legs for propulsion on land, moving like a snake. It could grow wings at will its head and back was covered with long hair.

Sisiutl appears to be no less strange associated with war, death and revival; it possessed supernatural traits, (such as being able to shape shift into a self propelled canoe! and having up to three heads or faces incorporated into its body, one being human like). The Indians however believed it to be a living, breathing creature. Depictions of these beasts from Eskimo and Indian representations, all seem to show common features, a serpentine body, crocodile like head and four to six limbs with a fluked tail. These graphic representations are usually adorned on local handicrafts and kayaks.

A significant, long and distinct neck is hard to identify from any of them and for an animal with a specific trait as reported by Mackal you would expect this to be a more prominent feature. My own enquiries to this region of the world have also so far been disappointing.

Following an email I sent to the Nunivak Island community I received a friendly but brief reply informing me that with regard to Palraiyuk the Eskimos did indeed use its symbolic form on their kayaks but were unsure of its origin. Tizheruk was not a known creature of this area and it was suggested that it may be a creature from further `north`. [4] I have subsequently emailed the King Island community but as of yet have not received a reply. The only supporting evidence in favour of Tizheruk appears to come from Heuvelmans who included an account from a Russian explorer, Otto Von Kotzebue in *"In The Wake of The Sea Serpents"*.

During his first voyage around the world in 1815-1818 Kotzebue passed through the Bering Straits. He describes how he met an American working for a company in the region there since 1795 who related a tale of how he along with some others, had been chased by a huge sea monster. The description of the creature was found to tally well with native Aleutian belief.

It was the shape of a `red` serpent, immensely long with a head that resembled a sea lion. It had two disproportionately large eyes and stretched its head far above the water, looking for prey. It did this on a couple of occasions, scaring the local otariids into the water as it did so. So here again we have a fearsome sounding creature, although we cannot say that it is normally so.

It appears to conform to traditional descriptions of the feared creatures from this region and appears similar to Tizheruk.

As far as this present work is concerned, if Dr Whites information was accurate and Mackals notion of a pinniped is correct then it throws up an interesting question. For in all the accounts of long necked animals that appear in this work, there is no evidence at any time of aggression on the part of the witnessed creatures. Just the opposite in fact, for the animals described seem to be quite the reverse, shy and timid, inspiring rather than terrifying. So if therefore, the accounts included in this book represent one species of seal with a long neck and Tizheruk/Palruiyuk also turns out to be the same sort of animal, then are we dealing here with another, different, species?

• When seals attack....

Of course animals *do* get aggressive at times, but usually there is some reason behind it. For instance when injured, threatened, or defending their young, With regard to seals, aggression is usually aimed at their own kind especially when mating time is upon them (apart from of course the leopard seal). There do seem to be exceptions to the rule, as a swimmer off Leigh on Sea in Essex found out in 2003 when a six-foot seal attacked him and left him with a broken foot for no apparent reason (seal attacks in Britain are virtually unknown). [5] More recently `sick` sea lions, suffering from domoic poisoning, a toxin produced by algae, were seen beaching themselves daily at Manhattan Pier Beach, California. On one particular occasion a sea lion charged at a local surfer biting through his wetsuit and into his thigh. [6] A lifeguard in Santa Barbara similarly received thirty stitches to a sea lion bite[7] while a 13 year old `boogie boarder` was chased by a sea lion that took a chunk out of his board[8]. More disturbingly a Cape Fur Seal in South Africa has recently bitten of a woman's nose for which she must undergo reconstructive surgery[9]. If Tizheruk does exist however and is a seal, then the aggressive behaviour, which it displays, makes me inclined to think that this could indeed be a separate species. It is also worth remembering that Alaska is a rather cold place to be and although relict saurians may have been warm blooded, it would probably be much more comfortable for a pinniped or a whale.

Fig.18 Palraiyuk depiction on kayak.

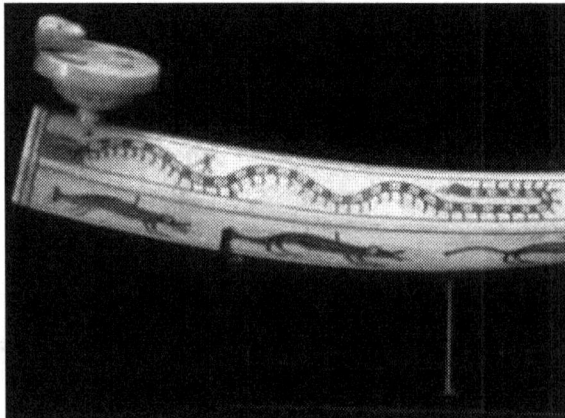

Fig .19 Palraiyuk (bottom) on decorative Eskimo pipe.

[1]. Pal-rai-yuk was also thought to have been a land creature at one time, living in marshes.[10]

6. CONVINCINGLY OBVIOUS?

So far, all the accounts we have studied are tantalisingly frustrating. On the whole they seem to describe very seal like creatures but although they could represent a long necked form of this species, it is also possible that they may be cases of mistaken identity, sightings of unusual or unfamiliar pinnipeds. Many people already have a mental picture of how a seal should look and when confronted with a pinniped such a lurching elephant seal or marauding leopard seal, may feel that they are indeed witnessing an unknown animal.

However, in this part of the work, we shall investigate reports that unless viewed as fabrications, seem to describe a presently unknown form of seal with a long neck, as mistaken identity does not seem to fit.

- **The Filey Encounter**

This British encounter with a Sea Serpent, seen on land, took place at night and was apparently reported in the *Daily Telegraph* for the 1/3/1934.

It was witnessed by Mr Wilkinson Herbert, a local Coast guard who had taken a wrong turn along the sands at Filey Brigg in Yorkshire during the course of his duty. Loch Ness had made the news the year before and the climate was right for such strangeness. Mr Herbert was walking along the beach, on a path near the waters edge, when he spied an object crawling over some black looking seaweed;

" Suddenly I heard a growling like a dozen dogs ahead, walking nearer I switched on my torch and was confronted by a huge neck, six yards in front of me, rearing up 8 ft. high! The head was a startling sight- huge eyes like saucers, glaring at me, the creatures mouth was a foot wide and neck would be a yard around. The monster appeared as startled as I was. Shining my torch along the ground I saw a body about 30ft. long. I thought this was no place for me and from a distance I threw stones at the creature. It moved away growling fiercely and I saw the huge black body had two humps on it and four short legs with huge flappers on them. I could not see any tail. It moved quickly, rolling from side to side, and went into the sea. From the cliff top I looked down and saw two eyes like torchlight's shining out to sea 300 yards away. It was a most gruesome and thrilling experience. I have seen big animals abroad, but nothing like this."

Although there have been some sightings of Sea Serpents from around Yorkshire, to the best of my knowledge, it is not renown for them and coming just the year after the Loch Ness affair broke, sceptics may well view it as some form of tourist ploy. Strangely, or coincidentally maybe, this was not the first time a strange water monster had been seen around Filey.

Sometime before or after this sighting local fishermen had been wondering why the fish had disappeared from the Filey coast and were experiencing their worst season in living memory when a similar creature was then observed in the area at sea.[1] One witness claimed it was bigger than a motorboat and had a `tree trunk` neck, while another described small eyes, a big head and a long *"raking body with two bumps on it"*.

And in fact a very long time before Mr Herbert's encounter, no less than a Dragon had been seen. [2]

- **The Filey Dragon**

The Filey dragon legend concerns a local Taylor, Billy Biter. Mr Biter was walking along the cliffs one mediaeval, misty morning, when he tumbled into a ravine that turned out to be the lair of a dragon. The dragon was about to devour him when Billy offered a local sweet delicacy, a parkin, which the dragon enjoyed so much, that he let Billy go. Billy told his wife who then set about baking the biggest and sickliest parkin in the whole of Yorkshire, which when offered to the dragon, caused its jaws to become stuck tight. The dragon then flew into the sea but could not overcome the icy waves. Its bones turned to stone and became Filey Brigg.

A truly delightful account with some similarities to Mr Herbert's Beast so did Mr Herbert feel the time was right for another dragon?

Doubtful, as apparently he received quite a bit of ridicule from his colleagues.

Anyway, thanks to his torch, we have a good basic description of a large, unidentified animal, with a distinct neck, four flippers, no obvious tail and one that seems to move like a sea lion.

The animal growls fiercely and has eyes ` *as big as saucers*` that shine `*like torch-light's*`, (seals have very reflective retinas, which reflect light in much the same manner as a cat). There is no description of fur or whiskers, but under the circumstances, this is understandable.

So what did he see?

The basic description would seem to fit a seal identity with some similarities between the creature seen and for instance a rearing Elephant Seal, (albeit a long way from home), which would also approach the length described. However, an Elephant Seal would not show the sort of neck that Herbert seems to describe, even if rearing and as we have seen would not roll from side to side when moving, being a phocid.

At face value then, a large otariid seal with a long neck would fit the bill rather nicely and although I feel this is a fairly convincing case for such a creature, there is one, very remote possibility concerning out of the way pinnipeds. It is one that the reader may feel should have been mentioned before with regard to mistaken identity.

For many years The Brisons, a group of rocks off Cornwall, have been the home to an extremely out of place pinniped, a Stellers sea lion (*Eumetopias jubatus*). Nobody is sure how such an animal, which is usually found in the North Pacific to the coasts of Japan and Mexico, got there, but it is a member of the largest species of otariid seal, which can reach 14 ft. in length.

So although an unlikely possibility as concerns the Filey account, a similar encounter of such an animal or a similar wandering species around the British coast could well account for eared, seal like sea serpents. In fact from what we have learned so far from the various descriptions, maybe there are regular otariid travellers to our shores.

Fig. 20   Steller's Sea Lion on
The Brisons   (Stephen Westcott).

A strikingly similar episode occurred in 1962 at Helensburgh on the Firth of Clyde, close to the entrance of Gare Loch. Jack Hay was walking his dog when it began to whimper and cower. Mr Hay later recounted:

"About 40 yards away I made out a massive bulk with a sort of luminous glow from the street lamps on the esplanade. It did not move for a minute, and then seemed to bound and slither into the water. I saw the thing swim out. It had a long body and neck and a head about three feet long. I watched until it was well out in the water and had disappeared. There was a strong pungent smell in the air".

Although scared he examined the spot where he had seen the creature and by the light of a match made out a footprint in the sand which had three pads and a spur to the rear. The sighting appeared to be linked with strange unexplainable noises that had been reported by residents and the fact that the canine population of the area seemed to be unusually reluctant to go out at night.

The luminescence he later explained was due to the street lamps which is slightly reassuring as otherwise it bares all of the hallmarks of a close encounter.

This bounding behemoth has many similarities to Wilkinson's monster but now leaves a nasty smell in the air.

As far as the footprints are concerned they do not seem to match anything normally found around British shores. If they were from a seal then you would expect them to be from a phocid as apart from wandering walruses and of course the sea lion at the Brisons, there are no native otariids. Phocid tracks tend to have a noticeable drag mark, where the animal drags itself along leaving paw prints on either side. Otariid tracks do not have such prominent drag marks as the animal's body is off the ground when moving.[3] They appear more semi-circular in nature and will also include a mark, sometimes indistinguishable, from their rear flippers which are drawn up to their fore flippers (could this give an impression of a spurred track?).

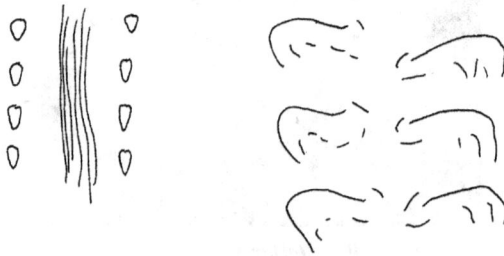

Fig. 21 Author's (poor) attempt at demonstrating phocid tracks (left) and otariid (right).

- **The Corinthian Encounter**

Although Heuvelmans includes the following account, it does not appear to have been widely published, probably owing to the highly romantic description given by the witness, G. Batchelor which could well be viewed as a modern `fairy tail`.

Batchelor apparently spied the creature whilst serving on The Corinthian when the ship was sailing off The Grand Banks of Newfoundland, in 1913.

I have included it in its full version here as found in "*In The Wake of The Sea Serpents*".

> " *As the `Corinthian` was ploughing her way westward, I was officer of the watch `on duty at the time`. At 4.30 a.m. in the cold gray dawn of August 30th, 1913, on The Grand Banks of Newfoundland, the look out man had just gone off and the third officer had left the bridge to see if all was well around the decks, while casting my eyes around the horizon I picked up an object*

about a mile off right ahead. The best conjecture I could make as to its nature was that it was a fishing boat lying end on to us. In the dense and extensive fogs, which sweep over the fishing banks sailors frequently become separated from their schooners and many starve for days before being picked up. I had just such an incident in mind as I watched the object ahead. When it suddenly disappeared beneath the surface, being still unenlightened I thought of tragedy. Suddenly however, after I had meditated upon serious things something surprising showed itself about two hundred feet away from the ship. First appeared a great head, long fin like ears and great blue eyes. The eyes were mild and liquid, with no indication of ferocity. Following sad eyes came a neck, it was a regular neck alright, all of twenty feet in length which greatly resemble a Giraffe. The monster took its time in emerging so long that I wondered what the end would be. The neck...seemed to be set on a ball bearing, so supple was it and so rhythmically did it sway while the large liquid blue eyes took in the ship with a surprised, injured and fearful stare. The creature was well fixed for side arms. Three horny fins surmounted its bony head, probably for defence and attack or for ripping things up. The body was about the same size as the neck very much like a monster seal or sea lion with short water smoothed fur. The tail was split into two fins. The colour scheme was good, although some might think it giddy; light brownish yellow tastefully spattered with spots of a darker hue. For a minute the creature inspected the `Corinthian` with its roving gaze, and then disappeared, showing its after works as it dived. Its whole attitude while in sight was that of one `moving about in worlds unrealised`. It seemed to be trying to comprehend a curiosity, which it had good reason to believe, might be a new danger. I almost felt tenderness for it and never have I experienced such a minute in my life. Down in my room I had a camera and a rifle. Yet I was the only one on the bridge besides the quartermaster at the wheel. I don't mind confessing that I wavered between my duty and desire for some sort of shot. Finally I stayed, but I don't know whether I should take full credit for that or not because I hated to lose sight of the thing. As it watched me it churned the water into foam and spray with its huge front fins. As it went out of sight it emitted a piercing wail like that of a baby. Its voice was altogether out of proportion to its size."

Wow!, what a charming account, liquid blue eyes and a twenty foot neck!

Now 4.30 a.m., on a cold `gray` morning, (not the best of times to see a sea serpent), awash in one of the roughest oceans of the world, it is easy to imagine a tired sailor glimpsing a seal, which to break the monotony of his watch and to keep him awake, takes on a Disney like reality, leaving him with a Grimm like fairy story to tell his grandchildren. In fact the reader may have already come to this conclusion, as clearly there is some degree of embellishment and high romance on Batchelor's part.

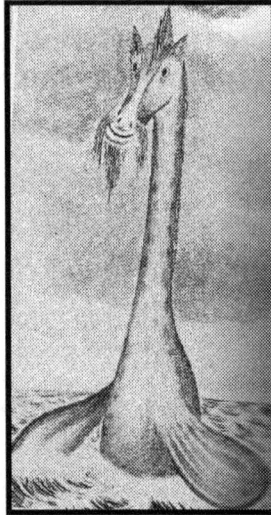

Fig.22 Batchelor's delightful sketch

However, Heuvelmans apparently had access to correspondence, which in his opinion validated the reliability of the witness. Batchelor's poor zoology, (horns for ripping) and the fact that the whiskers of the creature in his sketch, mirror his own moustache in a portrait that was also included in Heuvelmans book (an interesting bit of psychology?), did not deter Heuvelmans from accepting the account as genuine. There are in fact some genuine sounding details, which combined with the fact that Batchelor himself thought he had seen a plesiosaur, which he quite clearly had not, add credence to the account.

The climate of the time, predisposed to the notion of there being large saurians at large in the worlds oceans, yet despite this, Batchelor witnesses a mammal which instead of making a mighty saurian roar, wails like a baby. Matters are somewhat confused though, by Bachelor's assertion that such a creature may have played a part in the sinking of *The Titanic*! Now, what a film that would make.

From his drawing, the ears and eyes of the creature do not appear to be as big as he suggests and apart from the long neck of the creature and mane like fin, the sketch could be a highly stylised drawing of an otariid seal and the disturbance that the fore flippers make in the water suggests such an animal.

So did Bachelor see a known seal that stimulated his already obvious, vivid imagination, to produce a friendly dragon, or did he in fact witness a seal with a long neck? I think it is up to the reader to decide.

• The Discovery Island Encounter

This account comes from the waters of British Columbia and is one of many that have been attributed to `Caddy`, the resident sea serpent. It was reported by a professional fisherman, David Miller and took place in 1959, off the Discovery Light.

> " While engaged in commercial fishing... my partner Alfred Webb and I observed this strange creature surface roughly 80 ft. from our port beam. It started to move rapidly away from us so we speeded the engine up and gave chase. We got within 30ft. when it suddenly submerged, not in the method that seals and sea lions do, but as though something had pulled it under. A few minutes later we arrived at the place of submergence and there was turbulence suggesting a 30ft. sei. Whale. Its speed under water was also astounding as it surfaced a few minutes later over a hundred yards away. It stayed up while we took off after it again but this time it never let us get close again. The first encounter was so clear that both of us remarked about its large red eyes and short ears visible at that range."

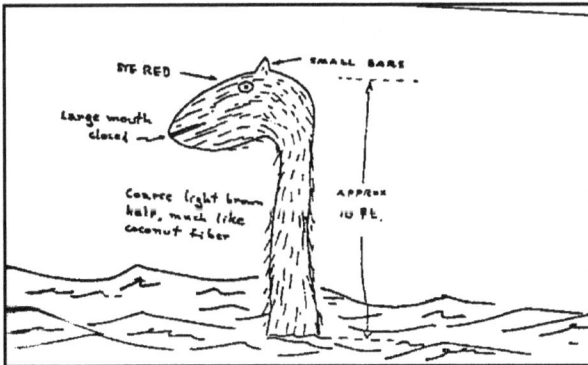

Fig.22 Discovery Island Creature.

This report is fairly straightforward and is excellently illustrated. Here we have a 10ft., fur covered head and neck, complete with small ears, characteristics that no presently known marine animal possesses and is reported by a witness who must be familiar with the local fauna. The head is very seal like and it is interesting to note that although the ears are visible, seen against the head of the creature in the sketch, they may only be distinguishable at close distance. Although the animal does not dive like a seal, surely the only obvious candidate for such a creature's identity, would be an otariid seal with a long neck.

- ## Briefly, Caddy

As briefly mentioned before, British Columbia has its own resident sea serpent, Caddy. Accounts of Caddy have been closely investigated by two scientists, P. LeBlond and J. Siebert, who initially concluded that three distinct, unknown animals, two of which that resembled Heuvelmans `Merhorse` and `Long Neck`, were probably involved. In fact other accounts of `Caddy` over the years have highlighted similar features to the Discovery Island one and seem to verify such a mammalian identity.

In 1934, two boys saw `Caddy` sporting a head like that of a cow with horns or ears, mounted on a 4ft. neck while a W. Kennedy spied a strange animal from his waterfront property in West Vancouver. It had a head that was serpentine in character, 18" length and 12" in cross section, without ears or horns. It was grey brown and smooth haired like a seal with a three and a half foot neck, which it held at an angle of 45 degrees.

Two large (40ft.) creatures were later seen in 1939, the largest one with hair on its horse like head and on its body while in 1950 a naval officer saw a creature with a head something like a garden snake, 18 inches across and two feet in length. It propelled itself with large flippers on either side and had a flat tail like a beavers (? hind flippers).

The head sat on a neck about six feet long and the head and body were covered with hair, brown in colour. More recently in 1984 a furry creature approximately 18-20 ft. long, 4ft. out of the water was seen with two large floppy ears and two stub-like horns by men fishing off the Spanish Banks in Vancouver.

Ears and hair? These must surely relate to a mammal, and what better mammal than a seal with a long neck?

Worth mentioning, before we move on, is a theory that has become popular in recent years for such sightings, especially for American Sea and Lake Monsters.

The theory holds that a primitive whale, something like the zeuglodon, may have survived to the present day and provided that evolution had not changed it significantly from its primitive form, hair and whiskers may still be present.

However, although such a theory is certainly not impossible as an explanation for some Sea Serpents, especially as the protecids, ancestors of the whales were thought to have fur, there is no evidence that any primitive whale had a long neck comparable with such sightings.

- ## The Mackintosh Bell Encounter

If this account is true and there appears no reason to think otherwise, then it is arguably the most convincing for the existence of a seal with a long neck. It took place off The Orkney Islands, (Hoy), in 1919 in the Pentland Firth, a body of water between the Orkneys and the coast of Scotland. It was not reported in the press of the time, but by personal correspondence to Rupert T. Gould who included it in is seminal work, *"The Case for the Sea Serpent"*.

Mr Mackintosh Bell, a writer or lawyer by profession from Moffat was on holiday helping out some friends on a local cod fishing boat. He planned to stay in the Orkneys for a month following recent demobilisation. It was his first day out at and around 9.30 a.m. on possibly the 5th August his friends had been telling him about a strange sea creature which they had seen on several previous occasions.

While inspecting Lobster creels, between Brims Ness and Tor Ness, Bells friends, who had commented on the animal moments before, pointed it out as it appeared, right on cue.

> " I looked and sure enough about 25-30 yards from the boat a long neck as thick as an Elephants fore leg, all rough looking like an Elephants hide, was sticking up. On top of this was the head which was much smaller in proportion, but of the same colour. The head was like that of dogs, coming sharp to the nose. The eye was black and small, and the whiskers were black. The neck, I should say, stuck 5-6ft., possibly more out of the water. The animal was very shy, and kept pushing its head up then pulling it down but never quite going out of sight. The body I could not see. Then it disappeared and I said `If it comes again I'll take a snapshot of it`. Sure enough it did come and I took as I thought a snap of it, but on looking at the camera shutter, I found it had closed owing to its being swollen, so I did not get a photo. I then said `I'll shoot it`, but the skipper would not hear of it in case I wounded it, and it might attack us. It disappeared and as was its custom swam close along side the boat about 10ft. down. We all saw it plainly, my friends remarking that they had seen it swimming just the same way after it had shown itself on the surface. My friends told me they had seen it in the same place the year before. It was a common occurrence so they said. That year was the last of several years in which they saw it annually. It did not show itself for two or three years, and then it was only seen once. As to its body, it was, as seen below the water, dark brown, getting slightly lighter as it got to the outer edge, then at the outer edge appeared to be almost grey. It had two paddles or fins at its side and two at its stern. My friends thought it would weigh 2 or 3 tons, some thinking 4-6. Not only my friends, but others lobster fishing, got many

chances of seeing it.

Dimensions; Neck, so far as seen, say 6-7ft. Body, never seen when neck straight up, but just covered in the water. You could detect the paddles causing the water to ripple. When underwater, swimming, the body, I think, to the end of the tail flappers would be about 12 feet long, and if the neck was stretched to 8 ft., the neck and body 18-20ft. long.

The skipper of the boat remarked that sometimes the top of the head, when seen from a boat vertically, was a bright red. Neck thickness say 1 foot diameter: Head very like a black retriever say 6" long by 4" broad. Whiskers black and short. Circumference of body Say10-11 ft., but this I am not sure of, as I never saw all round it, but it would be 4-5ft. across the back."

Bell's friends added rather mysteriously that if another hot summer prevailed, the creature would be seen again.

Bell himself included two excellent little sketches that seem to leave little doubt to what sort of animal we are dealing with here.

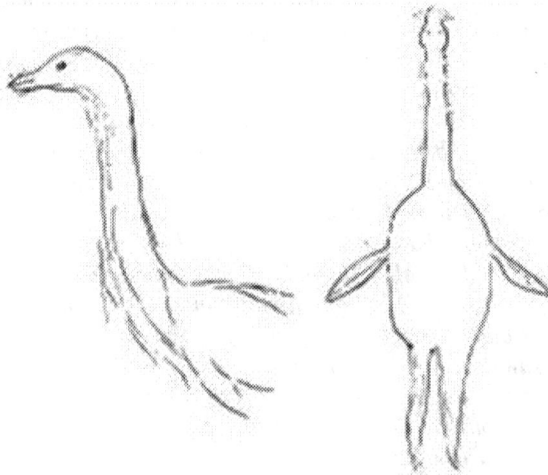

Fig.24 Mackintosh Bell's creature.

Bells sketches are pretty convincing and clearly denote some form of pinniped form that appears to have a long neck. The description and size is fairly small and such a size is not unknown from presently known pinnipeds, but no known species, could present such a long and well defined neck.

There do not appear to be ears present, from both the account and the sketches; however, in Heuvelmans version of the sketch there appears to be some attempt at illustrating something, on the profile of the animal's head. There is no tail apparent or described, just the rear flippers that could give a divided " tail" impression.

Now seals are common around the Orkneys and must surely be seen frequently by fishing boat crews, so perhaps the greatest mystery of this account is why didn't Bell's friends relate their sighting to such an animal species? Admittedly it does not look like a conventional seal having a long neck but even so apart from this every other feature of its morphology is in keeping with known seals even down to the whiskers. Although the creature seems hairy from the sketches there is no remark on fur as such. The `red` appearance sometimes seen on the top of the head is quite an unusual feature but to me makes this account more convincing as it appears to serve no purpose in relation to exaggerating the sighting.

Was Mackintosh Bell then, a clever prankster anticipating Heuvelmans by forty years, or did he and his friends observe an unknown species of seal with a long neck?

A few years before in 1910 a W.J.Hutchison, his father and cousin who were sailing for the Skerries to shoot duck and plover saw a similar type of creature. Their attention had been initially caught by a school of agitated whales moving away from the area at high speed when they observed a creature with a snake like neck and head much like that of a horse or camel. Hutchison also thought about taking a shot at the animal but was convinced by his father not to, lest the animal turn and attack them.

Nor was this the only previous time such a strange beast had been encountered.

Around 1850 a young boy, Alec Groundwater was spending a day in Orphir situated on the mainland. He was on the shore, perched on some rocks, gazing out over the Scapa Flow his legs dangling out into the water. The sea beneath him suddenly began to boil and an animal with a broad, flat head and a wide mouth containing some "wicked-looking" teeth or tusks surfaced. It apparently glared fiercely at Alex with "cold baleful eyes", before rearing up and attempting to seize the lad's dangling legs.

Fortunately the monsters attempts failed and it eventually plunged beneath the sea surfacing once to "shake its head and mane till the water cascaded from it on all sides, then disappeared".

- **Further Conclusions?**

The Filey encounter is frustrating. As it occurred at night further detailed information with regard to fur etc. is not available. The element of surprise also probably obscured events. As it happened at a time when Loch Ness would have been fresh news, there could be elements of exaggeration. However as a similar creature was seen in the locale at sea and the fact that Wilkinson's beast lacks a conspicuous tail, I think a hoax is not the case. A large wandering otariid similar to the Stellers sea lion at the Brisons may offer a remote possibility.

Batchelor's account is more difficult to accept when the fairy-tail features of his animal are taken at face value. The fact that he thinks he is watching some form of plesiosaur, which seems to have changed into a mammal somewhere along the line, would seem to indicate the account is true. Given his vivid imagination he could have clearly made a more convincing case for a reptilian leviathan.

The Discovery Island encounter is difficult to identify, if it's not a seal with a long neck. The animal has fur and ears. The sceptic will no doubt promote some form of temperature inversion, which would perhaps make the neck appear longer than it really was. The fact that similar creatures have been seen in the general area, sometimes when the weather is poor however would seem to indicate that the encounter was as described. The witness, being a professional fisherman, would surely have experience of such phenomena anyway.

This brings us to the Mackintosh Bell report, perhaps the most convincing.

There is little to be said here unless the encounter was a complete fabrication. We have a seal-like description and shape for the animal along with sober measurements and observations that you would probably expect from observing an animal so clearly.

It is a shame Bell's camera seized up. Heuvelmans could have been saved a lot of trouble.

As I have stated previously, this work is not intended to convince anyone that the Surreal Seal exists. It would be nice to gain support for future research and contemplation but ultimately I am trying to make sense of what some witnesses of sea serpents are actually seeing then trying to build a framework for the existence of a creature that best fits their descriptions.

The following part of this work will concentrate on doing just this.

PART THREE

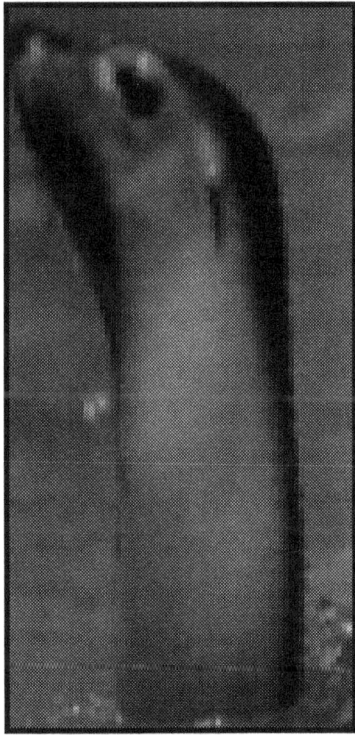

AN
UNDERSTANDING

7. QUESTIONS AND ANSWERS?

Having reviewed some of the witness accounts of long necked Sea Serpents, it may have become strikingly clear to the reader, that a seal with a long neck would offer a good solution as to their identity. However, up to recently there are two common arguments put forward against the existence of such a creature.

These are that firstly there was no evidence to suggest that a species of pinniped from the past or present, had ever evolved a neck comparable to that of witness descriptions and secondly that if such a creature did exist, being a pinniped, it would be tied to land and hence therefore discovered by now.

However, with regard to the first proposed pitfall, in the early 1980s, two new species of fossil pinniped were discovered in South America which appear to come close to such an identity.[1] Apparently belonging to the phocid family and dating from the Pliocene, these seals were characterised by their *elongated necks*.

Although there appears to be little available information with regard to them outside of academic circles the species *acrophoca longirostris* has been termed a `swan necked seal`.

Although in reality these would not be comparable to those of plesiosaurs, they are significantly longer than the necks in presently known pinnipeds. What is more there appear to be other as of yet undescribed members of the genus which may prove to share similar or more pronounced features. In fact one, *acrophoca piscophoca*, apparently has an even longer head and neck.

Recent research carried out by Dr Leslie Noe of Cambridge University has suggested that in fact plesiosaurs could not hold their necks above the water due to their osteology. Instead they used their necks as a feeding tube feeding along the bottom of the water.

The size of both these animals appears to be generally consistent with presently known species of pinniped and it was originally proposed that both were related to the modern leopard seal although there is a school of thought that places them closer to the monk and elephant seals.[2]

So with regard to a seal having evolved a long neck there is at least some tentative proof to show that two species of apparently-extinct pinniped appear to have done just that.

If these pinnipeds managed to survive to the present day, or the feature was adopted by a similar group of pinnipeds then such a feature must have some evolutionary use.

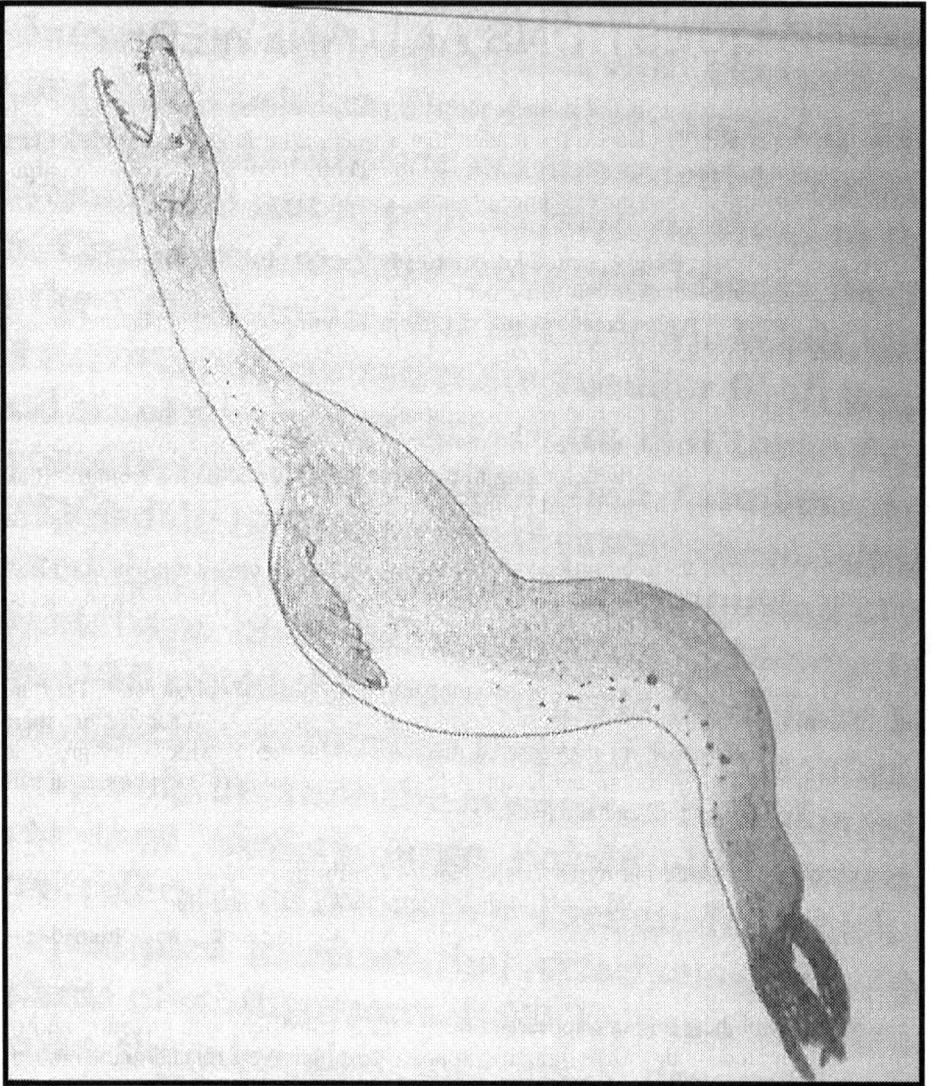

Figure 25. Reconstruction of Acrophoca Longirostris , D. Naish.

The second argument against such an animal surviving without modern detection, however, needs to be addressed.

In the next few pages I will attempt to offer speculative ideas, based on information known about existing pinnipeds, to deal with both these arguments and attempt to describe a theoretical natural history of such an animal.

- ## A LONG NECK?

When people think of a long neck they often think of a giraffe or Plesiosaur, both of which have been used to describe sea serpents. The giraffe, of course, uses its long neck to feed on the leaves of tall trees, while the plesiosaur used its neck to catch fish and possibly pluck things from the sky. In both these cases the neck is used for the primary need of any animal: to obtain food.

Unlike the cetaceans, whose necks have become compressed and fused for their aquatic adaption, the necks of pinnipeds are extremely flexible and when this attribute is combined with their swimming ability, they become fast, efficient predators. As well as krill, squid and fish, the diet of pinnipeds may also be supplemented by predation on birds and other small animals. Although not fully assessed in most species it is common in the Northern fur seal and unquestionably voracious in the leopard seal. Indeed the leopard seal is known to regularly gorge itself on penguins and other avian prey as well as other species of seal.

As we have already seen, it is also known for its very reptilian appearance and sinuous neck, which it actually uses in predation by coiling it back and striking at its prey like a snake. This method of predation seems to be similar to how it is thought that the plesiosaur must have hunted, by `darting` its neck out at passing fish.

In fact, Heuvelmans suggested that a seal with a long neck might have evolved to feel the niche left by the plesiosaur and its kin and given the descriptions of the leopard seal, it is not hard to accept such likelihood. So, theoretically at least then, a seal with a long neck may gain some advantage in predation, especially if it was able to conserve vital energy by utilising such an asset (something that I will speculate on later). However, a long neck would not have appeared overnight and would have had to evolve in response to a particular environmental need or function.

Now although otariids use their necks in locomotion, both on land and in the sea, a very long neck would surely hinder underwater efficiency in such an animal, as the ideal underwater form is a cigar shape, slightly tapering at both ends. As already mentioned the cetaceans have compressed necks, which provide them with a streamlined shape, and when phocid seals swim they tend to `telescope` their necks, thus shortening their bodies and reducing underwater drag. Now, I will not even attempt to take on underwater dynamics, as I am poorly qualified to do so; suffice to say that both the plesiosaur and apparently acrophoca longirostris, however well they swam, had elongated necks.

There are a few other areas that I would like to explore concerning the possible use of a long neck, although again I am not qualified to do so. I have included them in the hope that may inspire further discussion.

The first of these concerns thermoregulation, or the regulation of body heat. As seals are well insulated by fur and blubber, when the climate becomes warmer they may suffer from a form of heat stress. A common way of relieving such stress, for instance on land, is to rest in tide pools or seek rocky shelter. They also have a unique ability found in some marine mammals in that they can divert warm blood into their flippers, due to the close proximity of both veins and arteries in their bodies and the flippers can then be dipped in water or waved in the air to cool off. (Seals cannot sweat like humans, although the Northern fur seal can pant like a dog.)

Likewise, when pinnipeds need to conserve heat, species such as the California sea lion may adopt a `jugging` position in the water whereby their flippers are displayed on the surface of the water.

The animal will rest with one of its fore flippers and both of its hind flippers above the surface in a sort of basking position.

Would therefore such an extremity as a long neck, with the relevant type of circulatory properties, fulfil a similar function? For instance the seal could simply rest its head above the water to conserve or lose heat, and it is interesting to recall Mackintosh Bell's friends, alluding to the fact that the creature was seen when the weather was hot. Other points to possibly contemplate with regard to this feature are whether the long neck provides some mechanism whereby the seal can retain a greater oxygen carrying capacity that may allow it to dive for longer periods or stay submerged longer. Elephant seals have stomach sinuses that allow them to store more blood to do just this.[2]
Or could it also incorporate a biological structure which also allows a unique surfacing and submerging capability that may explain why many reports of sea serpents comment on the ability of such creatures to "sink" straight down? Possibly some structure akin to the pharyngeal pouches found in the walrus?

And what are we to make of the Storsjo animal's extending appendages? Although this account is difficult to make sense of and I am not convinced by Heuvelmans assertion of `snorkels`, which he envisioned would function by allowing his animal to breathe without surfacing fully, there is some relevance for the idea in the pinniped kingdom. As well as helping such a creature to breathe without surfacing fully while swimming, such an adaption may be of benefit in other ways. Seals such as the northern fur seal spend long times at sea and must therefore sleep in the water. When sufficient hauling out ground is not available, other species such as the harbour seal, take on a vertical position in the water whereby their body remains submerged, but their head pokes out above the surface. The elephant seal, the largest of the pinnipeds, has also been observed in this position and the walrus as already recounted as buoyancy pouches which allow it to sleep afloat.

So a long neck, equipped with Heuvelmans snorkels, may provide a novel method of not only breathing while swimming, but also breathing while sleeping and in both cases, reduce the need for being observed, also a good ploy in predation. Such breathing tubes are unknown in present pinnipeds, but two species, the elephant seal and the hooded seal, already have nasal protuberances.

The male elephant seal has its weird looking inflatable proboscis, while the male hooded seal *(Cystophora cristata)* can actually inflate a red, membranous sac, from its left nostril, (surreal indeed).

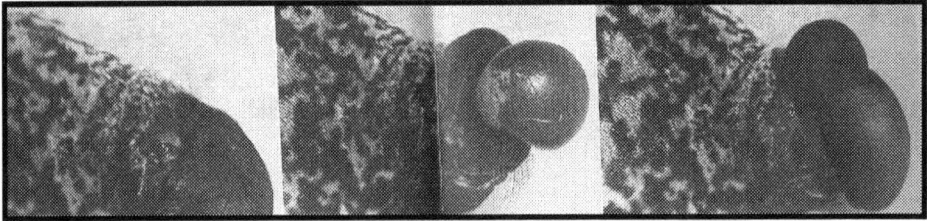

Fig. 26 Hooded Seal at various stages of hood inflation.

These unworldly assets fulfil roles in sexual and territorial display in both animals while in the elephant seal the proboscis helps in the conservation of water. In both cases such adaptive strangeness is only found in the males of the species. So although not impossible, such snorkels would be unlikely.

Still, it is worth imagining, a seal with a long neck, sitting submerged on a shallow tidal shelf, resting or sleeping with its head and neck barely touching the surface of the water, breathing silently through its snorkels while waiting for a tasty bird to set down on some nearby rocks.

After all, nature does provide many novel methods of survival for her creations. *

• SOCIABLY REPRODUCTIVE?

Although some species of seal may spend up to ten months at sea, they must all presently return to land for the purpose of moulting, breeding (although copulation may occur on land or in the water), resting and giving birth. Such hauling out time varies between species and depends on a number of factors including available space, site fidelity and the foraging patterns of individual species.

---

* *According to scientific principle the fineness ratio is used to predict the effect of underwater drag on a body shape. It is calculated by dividing the length of the swimming object by the width at its widest part. Marine mammals exhibit a value from 3-7 although 4.5 is considered the most effective.*

If we take Mackintosh Bells animal then with a reported length of 20 ft. and a width of 4-5 ft at is broadest we end up with just such a value, 4-5. Whether a long neck reduces this coefficient is another matter and one that I am not qualified to explore.

• MOULTING

Moulting which is more pronounced and exhaustive in phocids, requires a great deal of energy expenditure and need for rest. However, the Northern fur seal *(Callorhinus ursinus)* may take up to three years to fully achieve its moult, a time when it is not tied to land.

• GIVING BIRTH

Although Heuvelmans quotes Victor B. Scheffer as saying that an occasional walrus or phocid may be born in the sea and survive, aquatic birth in pinnipeds as a rule does not occur. Phocids are probably evolving towards this practice, as their pup dependency may last for as little as four weeks and this adaption would presumably be the next evolutionary step for their species.

Heuvelmans, to help his theory along, opted for a species of pinniped that had overcome this need, but before we go on to discuss this notion it is worth looking at some other, already practiced strategies.

Pinniped birth in most species is well understood, although it seems that little is known about the reproductive behaviour of the leopard seal and few leopard seal pups have ever been seen, (despite the fact that there are approximately 220,000 Leopard Seals thought to be in existence).

In general a female seal, will usually mate a few weeks after giving birth, the resulting implantation and gestation periods being delayed for up to eleven months, until the same time the following year when the seal will give birth again. Seals usually give birth to a single pup and as we have seen, weaning may take a short time in phocids, but longer in the other two families. Lactation in phocids can last from 4-50 days while in otariids and odobenids it can last from 4-36 months, although female walruses will apparently take their pups with them when foraging.

Some harbour seal females *(Phoca vitulina),* congregate in shallow water or along sandbanks to give birth and their pups are born with a sea going pelt. If the need arises they are capable of following their mother into the water five minutes after birth. The hooded seal meanwhile has the shortest weaning period of any mammal, four days.

Arctic ringed seals *(Phoca hispida)* breed in solitary family groups, with an adult male, female and pup comprising one unit while some pinnipeds, such as the lake living Baikal seal *(Phoca sibirica)* and the ringed Seal, give birth and raise their young in ice caves which act as secure environments.

There are of course many caves the world over and it is clear that in certain parts of the world, for instance around Great Britain, that caves, usually inaccessible to man, play an expanded role in the life of the native pinnipeds for various purposes.

The rarest of all presently known seals, Mediterranean monk seals *(Monachus monachus)*, spend much of their time in and around such caves, usually with submerged entrances. They use the shallow waters around them as a learning ground and kindergarten for their pups to swim thus in relative safety from the surrounding environment.

The Hawaiian monk seal *(Monachus shauinslandi)*, whose skeletal structure has remained the same for 15 million years and is endemic to the islands, which give them its name, only number between 1300-1500 animals. They are also solitary and are rarely found in groups preferring remote atolls away from humans.

Similarly, not much is known about the breeding distribution of the grey seals *(Halichoerus grypus)* that inhabit Iceland, except that they appear to breed in caves.
What is more, it is only when an abundant species of seal hauls out, that we see the familiar sights of crowded, interactive hordes as typified by the elephant seals and sea lions, but not all species of seal are abundant. As we have already seen, the Caribbean monk seal is thought to be extinct, while at present there are thought to be somewhere in the region of 500 Mediterranean monk seals, 1500 Hawaiian monk seals and 2500 Guadalupe fur seals *(Arctocephalus townsendii)* in existence.

In comparison there are approximately 145,000 California sea lions *(Zalophus californianus)*, 550-750,000 Southern elephant seals and over a million Northern fur seals. Logically then, we could presume that if our seal exists, it is rare and if it has adopted for instance a cave rearing and birthing strategy, then instances of hauling out may be few and far between as well as remote. In fact considering the amount of sea serpents seen over the years, there may have been more sightings of such creatures than there are long necked pinnipeds existing.

There is also of course the question of survival. First year survival in British grey seals, which are abundant, is low with only 54% of females surviving to breed. A rare long necked pinniped species could therefore similarly lose half of its number in the first year adding to its rarity. On top of this there is also the potential effect of disease on any population. Although in UK waters there are probably few natural predators of such a creature if it does exist in other areas of the world this may be more common.
Alternatively, we can opt for Heuvelmans notion of aquatic birth, which would make the existence of such a creature easier to conceal. In support of this theory it is worth noting that a similar species of mammal, the sea otter *(Enhydra lutris)*, has already managed this. The sea otter is a relatively new animal species, having evolved to its present form within the last 5-7 million years, a much shorter time than the pinnipeds. In spite of this fact though, it has managed to eclipse them, as it is able to give birth on land or in the sea.

So, theoretically then, a recently evolved species of pinniped could have reached a similar advancement.

Whether any of this is correct or not, hopefully the reader will begin to appreciate that there *is* some scope for the possibility that a, rare pinniped that has evolved or adopted some of the strategies already practised within its kind, may not be as observable as one would expect.

In the following pages, I would like to expand some of this speculation further possibly wildly, in an attempt to produce a speculative natural history of such an animal.

9. A SPECULATIVE NATURAL HISTORY OF THE LONG NECKED SEAL

Although it is possible for a new type of species to evolve sometimes quite suddenly, I have used the following speculation to illustrate how a seal may have evolved a long neck and offered some scope for its theoretical evolution to its possible present day status. It is certainly not to be accepted as definitive as it is pure conjecture on my part. It is cited here in the hope that it will encourage constructive thought on the subject.

> *"About 5-7 million years ago, a species of seal, possibly otariid, began to forage inland traversing rivers and lakes. Such foraging activity allowed the species to exploit the abundance of food sources found in such areas and offered an ideal place to give birth and rear young."*

- **Evolution**

As we have already seen in the case of the sea otter, 5-7 million years is a suitable period of time for a new species to evolve and geologically speaking, is quite recent. We have some knowledge of pinniped dispersal in the past and presumably groups of pinnipeds would have broken off and explored different areas in search of suitable habitats, eventually leading to their present distributions. A relict population of harbour seals inhabits a Canadian lake, Lac du Loups Marins, near Quebec where for some reason they became separated from their kin about 8000 years ago. We also know that one species, the baikal seal adapted to a freshwater environment and has flourished to the present day.

I have chosen an otariid identity as otariids have a tendency for a longer neck and display more sexual dimorhism. We have learned however that a phocid such as the leopard seal also has a predisposition for a sinuous neck and that the fossil seals *acrophoca longirostris* and *acrophoca piscophoca* are phocine.

- **Passing Through**

Even today, species of seal, including the largest, somehow mange to overcome the many man made obstacles that they encounter to similarly make such journeys into lakes and rivers sometimes being found many miles from the sea. In the not so distant past, such obstacles would not have been present and therefore such a journey would have been easier to make. An otariid seal would also have the ability to be able to travel overland to a certain degree without too much difficulty. For instance, according to Mariane Reidman in, *The Pinnipeds*, a male California sea lion made two journeys from the sea into down town San Francisco and was found on both occasions in a public toilet! Apparently the Gents! Another group of sea lions apparently commandeered an old house on Ano Nuevo Island in California, climbing the stairs and finding solace in the bathtub. In fact otariid seals are very good climbers and can traverse large rock formations in their usual environment. Recently at a safari park in Wiltshire (UK) some have been seen literally `branching out`. A group of sea lion pups have been nicknamed `tree lions`, as they have been observed climbing along the limbs of trees that overhang a lake and diving off into the water.[1]

- ## Environmental Exploitation

At certain times of the year, for instance at spawning season, large numbers of fish would be found in rivers and lakes, offering an easy opportunity for predation. As we have seen previously some seals manage to take advantage of inland fish supplies found in Scottish Lochs while some California sea lions are actively practicing such exploitation, with groups found lying in wait at the mouths of rivers earning a reputation for being pests.

Riverbanks and the shores of lakes, may also have potentially offered plenty of room for hauling out and therefore giving birth which could have been timed to coincide with such a journey inland. The enclosed environment would possibly then be used as a convenient safe, nursery and it is known that different species of pinniped in different locations tend to exhibit patterns of movement and diving that suit their particular foraging habitat.

In June 1937 while out on a boat in Loch Ness, Anthony Considine and a friend reported that they had both observed three creatures off their stern, which were about three feet long. The creatures had long necks, four limbs and were swimming away from the boat. The witnesses likened the limbs to flippers and described how the rear two were held close to the body and were being used to push the animals forward. How delightful baby Nessies. Needless to say the report has been interpreted as being nothing more than a report of otters as three feet is about the size of an otter. The apparent lack of tails and method of movement does however remind one of a phocid seal swimming. If we also remember the Connemara pools, home to the elusive `horse eels` we could even speculate that the geography of this location is used as some form of unsupervised nursery. The legendary animals being reported here possibly are nothing more than juveniles, which could then account for the differences in description between them and other familiar long necked lake monsters.

Intriguing, possibilities, but just possibilities all the same.

At mating season most seals will starve, the females looking after their young, while the males compete for harems, which can be a very exhausting period. An enclosed environment may have therefore made it easier to obtain food during this period keeping the family unit together and once young had been reared, a return journey could then be made back to the sea.

"Over the centuries, these migratory journeys became an important part of the seals ecology, reducing the need to rely on a purely marine existence, thus enabling the species to thrive free from competition and predation. The rich and easily obtainable food supply, present at certain times of the year, promoted an increase in size, which in turn precipitated a need to conserve energy expenditure

As a result, the species began to rely more heavily on the flexibility of their neck to obtain food and in time the already flexible otariid neck grew longer, thus favouring energy conservation."

• Adaption

As we have seen, a pinniped, adapted to a marine existence, would have no difficulty in obtaining food in rivers and lakes, in fact, compared to the sea it would surely be much easier. A group of harbour seals apparently appears to practice this strategy now by occupying Lake Iliammna in Alaska during an all year stay. When species of pinniped are in direct ecological competition with each other, different species tend to specialise in a particular food source and a past pinniped species may have found this an easier option.

An abundant food source would presumably, have over time, lead to an increase in size and a subsequent increase in size may mean that there would be a need to conserve energy and adapt metabolism.

However, as the migratory environment was in comparison to the sea, enclosed, predation would be easier and a long neck to meet this need could conceivably reduce the need to conserve energy. It may also enable such an animal to exploit new food resources, such as birds or other shore living animals. What would be more useful to an animal trying to catch spawning and jumping fish than a long neck?

We would indeed have a mammalian Plesiosaur.

Paul Harrison, in his book *Sea Serpents and Lake Monsters of The British Isles* recounts correspondence with a lady called Helen Hadley who passed onto him an interesting report from some of her friends. (I had also corresponded with her some time ago). Apparently while kayaking around the Isles of Jura in Scotland her friends entered an underwater cave where to their horror they found the carcasses of several dead sheep. Whether these sheep had been sea-going feeders that had drowned or whether they were the victims of something far more sinister is not known. However if a large carnivorous animal has been forced to look for alternative food sources, well...who knows?

A head situated on a long neck may also have aided in navigation, both at sea and on the land and as we have seen, the neck in an otariid also enables it to move quite quickly when out of the water providing momentum by a swinging action.

> " As mankind grew in numbers, spreading slowly around the world, sightings of such creatures became embroiled in superstition and fable, giving birth to the `water kelpie` and similar folkloric tales. Continued exploitation of these waterways destroyed much delicate ecology along the way.
>
> From the dawn of industrialisation onward, mans ability for such, grew ever more impressive until today, whether it is lakes, rivers or the sea, mankind pursues a living and pleasure from bodies of water.
>
> As a result the species was forced to retreat into more remote areas and back into the sea, where it found itself in direct competition with its sea going cousins who were also competing against man, thus leading to a decline in the species as a whole. Today, if not already extinct or on the verge of extinction, such animals are rare, their infrequent appearances still generating tales of sea and lake monsters the world over. "

Here we have an interesting scenario, which could account for some lake monster as well as sea serpent sightings and may also go some way to explain their rarity today. In the last few hundred years mankind has spread and industrialised like no other time in its history with many adverse environmental effects, (which incidentally also seems to coincide with the rise of sea serpent reports although also with mans modernised assault on the sea).

It is not hard to understand especially in the present `eco friendly` climate how environmental changes can cause the demise of a specific species.

A rather old tale, which may be just that, comes from around Washington Lake in America. It concerns a local trapper who discovered the carcass of a miniature lake monster. It had a barrel like body, short stubby legs with webbed feet, no tail and a neck that looked like a snake. The head was small but contained a number of well-formed teeth. Unfortunately after seeking advice from a priest who thought it was the work of the devil, the body was burned!

9. PUTTING IT ALL TOGETHER

Although at the present time it is not possible to provide a completely accurate picture of the Seal Serpent if it exists, it is possible based on eye witness testimony and that which we have discussed so far, to attempt a rough description of such an animal and its possible lifestyle and characteristics.

So here goes.

- **SIZE**

Our creature is big, that much we can ascertain.

Apart from the Corinthian and the Noreen report, witnesses seem to describe an animal that varies in length from 20-30ft, quite a size. The largest of the known pinnipeds, is the elephant seal and the largest recorded specimen was a rather staggering 23 ft. in length. It is not until you compare such a size to a human however, that you begin to appreciate just how big this is. If we now add say, an extra 6 ft, for a long neck, we begin to approach a length of 30ft a respectable size for any sea serpent.

Anyone confronted by either beast, would I am sure find the experience a little unnerving to say the least and the sheer size and bulk of such a creature along with the element of surprise, may distort length estimates. An Elephant Seal lumbering down a beach, with its fat and blubber rolling in a rather nauseating fashion could easily appear a monster. Size may also alter at various stages of such an animals life cycle, for instance at times of fasting, lactation and foraging.

Fig. 27 `Homer`, an errant juvenile male Elelphant Seal.

Fig. 28 Scientist being chased by an Elephant Seal.

• HEAD

Batchelor describes a `giraffe` like head, although I personally think it looks more like a horse, while the `Discovery Island` animal appears to have a more seal like head. We cannot be sure of the Filey encounter, while Bell compares the head of his creature, which appears very small from the dimensions he gives, as dog like. Stoquelers creature and the Tasmanian beast also have dog like heads. This is interesting as there is apparently little difference between the skull of a dog and that of an otariid seal at first glance.

Some species of seal such as the Northern fur seal do have very small-looking heads, which seem to be totally out of proportion to their bodies. They also have more pointed heads than other otariids.

Fig. 29 Bull Northern Fur Seal, (Rolph Ream NMML 1992)

Seals, being mammals, would vary in appearance with age and the reader may have noticed the many different types of pinniped head, from the photographs included so far in this work. Here are a few more for reference.

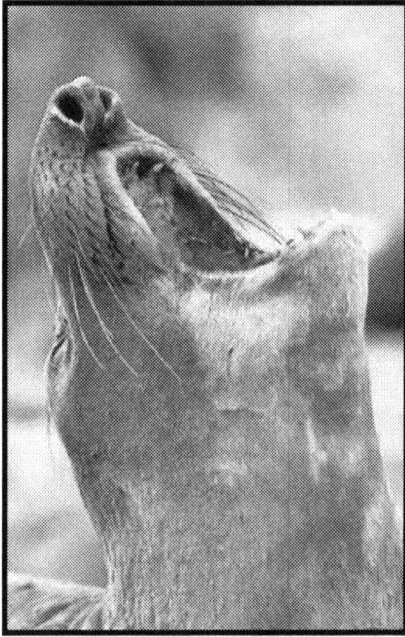

Fig. 30 Southern Elephant Seal (Karen French)

Figure 31 Crabeater Seals
(Dr J. Bergston, NNML)

Fig. 32 Male Elephant Seal (Karen French)

Fig.33 Stellers Sea Lion Swimming
(R.Ream NNML)

If our animal were otariid, then ears would be present as they are in the Corinthian and Discovery Island encounters. They may appear small and `clipped` looking, while when they are sleeked back, they may not be as obvious, e.g.? Bells report. When raised they may appear to resemble small horns, which may obviously confuse things, while if Heuvelmans snorkels are present, they are almost certainly inflatable, as in the elephant and hooded seal and therefore not visible at all times.

With the exception of the Filey account, in which the creature had a torch shone directly at it, the eyes of our animal do not appear to be much different from known pinnipeds, although there may be differences in size due to sexual dimorphism.

Seals also have whiskers, which are readily apparent in both Bell's sketch and Batchelor's drawing, but when seen from a distance they may not be as noticeable.

• NECK

The neck of our creature is of course its main feature and various lengths have been given for it, ranging from 6-20ft, (although I think we should take Batchelors account, with a pinch of sea salt!). Personally though, I feel that it may not be as long as the accounts may suggest, at least not normally. In the pictures below, a sea lion, when extending its neck, gives it a more pronounced appearance.

Therefore a six-foot neck may appear to be an eight foot one, if the animal was stretching for swimming or observation.

Another interesting speculation, in keeping with what is known about present pinniped species, is that some species of male seal for instance the elephant seal and male Otariids, tend to have a more pronounced head and neck region compared to females.

Therefore a female of the species may have a shorter, less conspicuous neck than the male. The mature male may also have a mane on its neck and back in keeping with known otariids such as fur seals. *

Both the South American Fur Seal and sea lion have visibly noticeable manes which are longer on the neck and shoulders than the rest of the body and in the case of the South American sea lion this also extends to the face, chin and between the eyes.

On a long neck as previously discussed, this may be more pronounced. The length, width and height of the neck may therefore vary, according to age, sex and stance.

---

* *It has also been speculated that reports of manes on sea serpents may in fact be a filament type substance such as that found on the Hairy Frog (Trichobatrachus), which in fact supplements pulmonary respiration. This theory, although unlikely and unknown in any other creature has been used in support of a non mammalian identity for such long necked animals.*

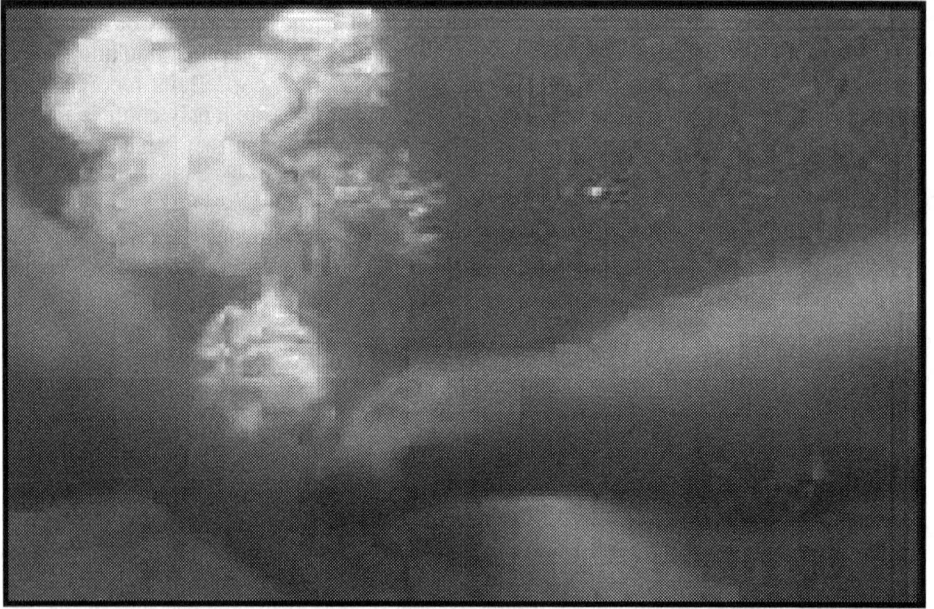

Fig. 34. Stellers Sea Lions Underwater(R. Ream NNML)

Fig. 35 Angry Mother Seal (NOAA)

- **BODY**

The body is probably massive and bulky on land, much like the elephant seal and as such would be covered in rolls of fat and blubber which may appear to ripple or become more streamlined when the creature is in the water.

In most of the accounts that we have studied fur is noticeable, although it seems to be lacking in Bells report and obviously the Filey account. All pinnipeds have fur covering the whole body, which may look water smoothed when wet and rough-looking when dry. The walrus however is sparsely haired and loses most of this for a two-month period during its moult. Moulting could also explain some of the discrepancies reported in colour, for when Hawaiian monk seals have been at sea for some time they may become covered in algae, which gives their fur a green-looking tint. *

The fore flippers appear to be fairly big, in keeping with otariids, while the hind flippers may appear tail like when splayed, or when the animal is stretched out.

Fig. 36 Mediterranean Monk Seal

- **DISTRIBUTION**

Heuvelmans carried out a geographical analysis of `Long Neck` reports, which showed what appeared to be a cosmopolitan range for the creature. There was a propensity of

Fig. 37 Stellers Sea Lions diving (note tail-like hind flippers).

reports from around the north coast of America, United Kingdom and Scandinavia in the northern hemisphere and from around Australasia in the southern hemisphere.
As a result he concluded that such an animal was a true cosmopolitan species and great traveller.

However, I think it is worth considering, if we accept the reports from Australia, that in keeping with presently known species of pinniped it would make more sense for there to be two species, such as for instance the elephant seals, a northern and southern one.
This would mean though, that either a group of pinnipeds evolved their long necks in one area before undergoing a species divergence, or that two species underwent a process of convergent evolution.

Also worth thinking about is Mackal's speculation on a northern counterpart for the leopard seal, for if such an animal is represented by Mackal's information then it seems to have a totally different disposition to the true `Long Neck`. This would probably make a sort of reverse logic if the traditional long necked sea serpent were otariid as if a phocid species, technically the most advanced, had evolved a long neck it may have then been mirrored in an otariid species. However, three new species of seal is probably pushing it a bit!

Personal instinct makes me more inclined to feel that a distribution pattern in the Northern Hemisphere is more likely. Probably ranging from the coasts of North America to the north of Ireland and extending to the north coasts of England and Scotland as well as Scandinavia.

However the actual range of habitation is probably much smaller, the wide distribution of sightings probably due to such animals foraging habits.

- ## HABITAT

If our creature exists and is rare, then its habitat may be little known. There are still many areas of vast, remote coastline around the world, which are sparsely inhabited.
If the species has also suffered ecologically in recent years, then it may have adapted a strategy similar to the Mediterranean monk seals, which due to environmental pressure have retreated from sandy and rocky beaches to submerged caves.

Stephn Westcott, a naturalist, who has been studying the seals in the United Kingdom for many years, reports in his book, "*The Seals of the West Country*", at length, on how the grey seals utilise caves for many aspects of their lives. Otariids also utilise such environments and in fact off of the east coast of America in Oregon, both fur seals and Steller sea lions share a massive cave system known as the Sea Lion Caves, the largest sea cave system in the world. Here they spend long periods at certain times of the year for instance in spring and summer when they breed and leave their young on rock ledges, just outside the cave.

Therefore a rare animal, utilising caves for existence would be infrequently seen, much like the Mediterranean monk seal and is it a coincidence that Heuvelmans geographic analysis coincides with areas of rugged, rocky coastline?

Figure 38, as an example illustrates the underwater caves present in the United Kingdom.

If you then compare the distribution of the UK's most common pinnipeds, the grey and harbour seals, their distribution would more or less correspond with the same areas. Interestingly - although statistically fallible - is the correlation between these areas and reports of sea serpents in the UK.

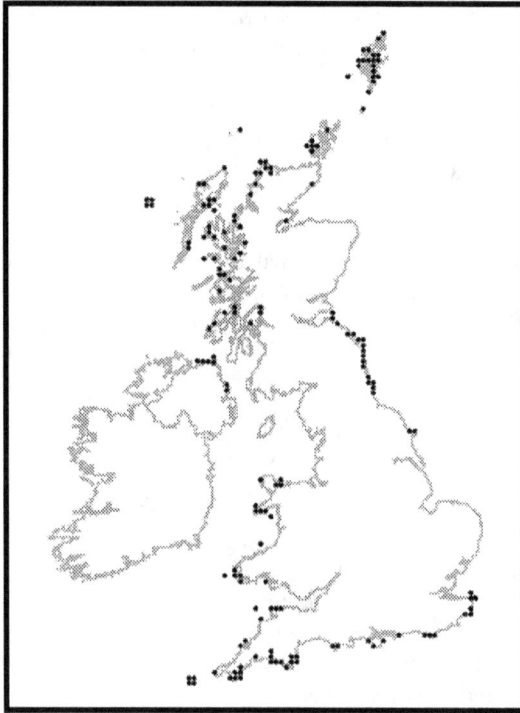

Fig. 38 Underwater Caves of the British Isles (JNCC).

The possibility of cave inhabitation may also be something that is universal when dealing with such a species.
Presently known seal species such as the fur seals do in fact prefer such rocky areas and what is more some such as the northern fur seal are very nocturnal in lifestyle, foraging, diving and hunting, mostly at night. Therefore if a long necked animal is rare, nocturnal and utilises caves it can be assumed that sightings may indeed be few.

As a rare species, our creature may also only exist in small, scattered family groups, migrating far and wide for various periods during the year.

The evolution of a long neck may also allow a more varied diet and if used in a role of energy conservation, may mean that such an animal is less likely to need long periods of rest, which would presumably reduce the need for land excursion.

Like other pinnipeds, such creatures would be well able to travel up rivers and to a certain degree and overland in search of food.

The accounts that have been discussed within this work are few and specific.

However if the reader accepts some of the possibilities that I have suggested then further research into these areas may be of particular interest.

• A QUESTION OF BELONGING

Having got this far, if we are to accept the possibility that a species of seal has evolved a long neck, then we are obliged to try and classify it within the currently known families.

Without physical evidence this is impossible, but if we use the eyewitness accounts previously described, we may be able to draw some conclusions.

• LONG NECKED SEAL AS PHOCID

The phocids are the most advanced of all the pinnipeds.

If we accept Heuvelmans notion of advanced aquatic adaption then it should be within this classification that our animal belongs. *Acrophoca longirostris* and *piscophoca* it should be noted were both phocine. However a neck as long as accounts would suggest may hinder aquatic performance and as the phocids are evolving more towards a totally aquatic existence this feature seems slightly at odds. Was acrophoca logirostris then an evolutionary dead end or did a phocid adapt a form that filled a certain marine niche?

The other potential problem is, that as illustrated in the *Corinthian* and Discovery Island accounts, ears seem to be present which is predominantly an otariid trait although again phocids such as the harbour seal may develop external ear pinnae.

• LONG NECKED SEAL AS OTARIID

From the point of view of ears and locomotion and if we accept the Tasmanian and Filey encounters as true `unknowns`, this classification would make more sense and there would probably be more room for discrepancy in description, due to sexual dimorphism. Fur would probably be more noticeable in an otariids species although in Bells account it is not obviously apparent, although the animal could have been in moult.

No otariid inhabits the North Atlantic and a wide foraging dispersal from this area both east and west may tie in with sea serpent reports.

If we wish to accept the Heuvelmans theory of adaption and birth at sea though, it may have to be a new otariid species, one which has managed to surpass the phocids in terms of specialisation.

- **LONG NECKED SEAL AS ODOBENID**

I think it is fairly safe to conclude that there are not species of long necked walrus swimming the oceans, (modern tusked ones that is) but it should be remembered that this family possess both phocid and otariid characteristics. Interestingly, fossil Walrus, of which there were once many different species, do not seem to have possessed tusks, implying that this attribute is only a *recent adaption* and as already mentioned there were many tuskless species in the past.

Odobonids also display both phocid and otariid traits and an odobenid that had developed a long neck instead of tusks is an interesting supposition.

- **LONG NECKED SEAL AS NEW FAMILY ORDER**

Here we could imply that a recent branching in pinniped evolution has resulted in a new classification of pinniped family, which like the odobenids shares both phocid and otariid characteristics as well as having been able to evolve a more marine existence perhaps enabling it to give birth at sea. As we have seen in the case of the sea otter, this is not as unusual as it may sound and could have taken place in a relatively short geological time. While this categorisation is quite attractive, it seems a little bit of an easy option.

My own feeling is that the animal if it does exist, shares more otariid traits than phocid or odobenid, although it may not actually have to be an otariid to do this. No otariid species are native to the North Atlantic. How the evolution of such a creature would fit into the modern theory of pinniped evolution though I do not know. However it would probably cause quite an unwanted stir.

10. CONCLUSIONS

If we accept the hundreds of Sea Serpent accounts over the years and from all around the world which have been reported it is pretty obvious that they are not due to just one species of possibly unknown animal.

There are too many differences in size, shape, colour, behaviour and appearance for this to be the case. There is probably no doubt a few accounts have been of rare, unfamiliar, out of the way animals, while others are probably cases of plain mistaken identity. Yet on the whole they appear to imply a body of evidence that indicates some large presently unknown animal existing. With regard to the proposed existence of a long necked seal the reader by now, may have already come to his or her own opinion. Without hard proof it may be hard to contemplate and more difficult to accept.

On one hand, common sense would imply that by the law of averages, a large unidentified seal with a long neck, because of its pinniped nature, would surely have been discovered by now whereas on the other, if such a creature does exist and has adopted some of the speculative methods of existence that I have discussed, then the odds against its existence may be lowered somewhat.

The intention of this work has never been to prove beyond doubt that such creatures do exist, but has focused more on the possibility that they *could* by (hopefully) adding some thoughtful insight on the subject.

It is obvious to any reader however, that people reported sightings, which seem in some cases to promote a mammalian identity. If these are genuine and we accept this then we are left with several choices.

- **A totally new, previously unknown form of animal species has evolved;**

- **That a species of plesiosaur survived extinction, (possible?), became warm blooded, (not impossible), gave birth to live young, (absence of eggs, again possible) grew fur, ears and whiskers, turning into a mammal along the way;**

- **a species of seal grew a long neck (known).**

A fourth option that could be relevant to UK sightings is that rogue populations of known seals such as the fur seals or even the elephant seal are inhabiting places that are normally out of their distribution pattern (Scandinavia, Scotland?).

Whether such seals could then go on and give the appearance of a long flowing mane as reported in some instances of sea serpents however is doubtful.

Similarly it would not be surprising to find that that various existing pinnipeds travel further and wider than is presently known, turning up in lakes and rivers tremendous distances from their normal habitat. Being unfamiliar to the populations that reside there they may subsequently cause confusion and misidentification as well as perhaps wishful thinking on the part of any witnesses who subsequently observe them.

In the vast expanse of water on this planet, there is plenty of room for plesiosaurs, long necked seals and even updated zeuglodonts although similar looking creatures, sharing a similar marine niche, is unlikely.

- **Physical Proof**

For the above scenarios, there is of course a total lack of physical evidence, something that in today's world full of scientists studying animals in far-flung reaches of the globe and people owning video cameras is rather surprising. Conversely however, science still does not recognise `the sea serpent` and is not actively involved in searching for either a plesiosaur or seal with a long neck, let alone listening to the accounts of reliable witnesses, or speculative support for such notions.

When enigmatic carcasses turn up on the shores of beaches they inevitably turn out to be strikingly reptilian, decaying basking sharks. There are probably many bodies washed ashore or found on beaches however that are simply left to rot or are not equated with anything unusual.

Lack of physical evidence may not be that hard to accept bearing in mind that pinnipeds, swallow stones, which on death may cause their bodies to sink without trace. It is not known for sure why they swallow stones but may be due to a need for ballast, a food substitute or for grinding internal parasites up and it would be interesting to learn just how many dead pinnipeds are washed up after scavenging nature has taken its course.

There is of course the possibility that such creatures as sea serpents and other mysterious animals are not physical at all, materialising through that still little understood portal of the human mind.

This something that some researchers are considering or concluding.

Although this may be a valid solution, it is one that I would prefer to ignore for the time being.

- **The Surreal Seal Question**

As we have seen, people the world over appear to have witnessed animals that seem to resemble prehistoric reptiles, the foundations for such identities being layed long ago, popularised by the media and of course Loch Ness. This popular belief may indeed have affected the discovery of `The Surreal Seal`.

Everyone knows what a plesiosaur looks like and most people are familiar with such an animal and its physical appearance. Therefore if a witness sees a long necked sea serpent, it is almost inevitable that they will correlate it with a reptilian culprit such as the plesiosaur, for how many people are familiar with the theory for a seal with a long neck? In recent years, sightings of such creatures seem to have dwindled, (correlating nicely with the theory of threatened ecologies), but such animals if they exist must still be being seen. Could it be that in the technophobic twenty first century, a time of computers and the Internet, people are willing to ignore what they see, as they glance away from their laptop? Or is it that in such a technologically advanced age, there is no room for the romantic notion of The Sea Serpent?

It would come as no surprise to learn that people are still seeing such creatures but are either reluctant to speak of such encounters for fear of ridicule, or that the areas that such creatures must now inhabit are remote, populated by small communities who simply accept such creatures as part of every day life. If this is the case then they are doing the animals a favour, unconsciously protecting them from possible ecological or curious catastrophe. Whatever the truth, there is a place for The Surreal Seal to exist, just as there is a place for the plesiosaur and others.

Possibly more so considering some of the evidence in this work.

It is on this note that I would like to finish this speculative journey. I hope that I have updated the notion of a long necked pinniped and have raised some interesting speculation that has not been offered before. Some of my speculation will no doubt be faulted and I ask the more scientifically grounded reader to excuse such amateur shortcomings. However at least I have dared to contemplate.

By doing this I hope that I have made the notion for the existence of The Surreal Seal more acceptable, at least in theoretical terms.

FOOTNOTE: *Anyone who is interested in furthering the notion of The Surreal Seal, or wishes to go in search of such an elusive but charming creature, is more than welcome to contact me and keep my head filled with wonder. robert@cornes1.fsnet.co.uk*

APPENDIX 1
THE FLY SEA SERPENT

The following account has often been published with a delightful drawing of what looks very much like a seal with a long neck.[1]

However, before we go into this, we will examine the account. The sighting occurred in 1849, and was made by a Captain or Lieutenant George Hope, whilst sailing around The Gulf of California, aboard *H.M.S. Fly*. The account was told in company and was related by a lucky guest, to Edward Neuman, a naturalist who published it in *The Zoologist* of 1849.

> "... he saw at the bottom a large marine animal with the head and general figure of an alligator, except that the neck was much longer, and that instead of legs, the creature had four large flappers, somewhat like those of a turtle, the anterior pair being larger than the posterior; the creature was distinctly visible and all its movements could be observed with ease. It appeared to be pursuing its prey at the bottom of the sea, its movements were somewhat serpentine, and an appearance of annulations or ring like divisions of the body were distinctly perceptible."

Fig. 39 H.M.S. Fly Sea Serpent. (C.M.Park)

The drawing is delightful.

Unfortunately, from all so far printed referrals to this account that have appeared, none makes it clear if this sketch was based on further deliberation with the good captain, or whether Carton Moore Park, used his own imagination to fill in the gaps.

Considering the sketch, and the area in which the sighting occurred, one could draw similarities with a Californian Sea Lion. In fact, even in the original article, a copy of which I have viewed, there is no further reference to such deliberation and therefore, despite being a rather intriguing sketch, complete with mane, whiskers and some discrepancy for a tail, it must only remain a true unknown.

SELECTED REFERENCES

Chapter 2.

1. "Seals in Shetland"; age of oldest seal, wildlife.shetland.co.uk/marine/seals.html

Chapter 3.

1. "The Bunyip"; Gary Opit, Myths and Monsters Conference Papers 2001.
2. "The Bunyip"; Moreton Bay Free Press , kindly supplied by Pam Cory of The Brisbane Historic Society. www.brisbanehistory.asn.au
3. Schlemman 1990,"*Crizmeks Encyclopaedia of Mammals*", Vol. IV, New York McGraw Hill Publishing Co.
4. See reference 1.
5. See reference 1.
6. "The Suffolk Sea Serpent"; *Animals & Men*, Issue 33. p18-19. Centre for Fortean Zoology (CFZ), as well as personal email and telephone correspondence with Mr Picard.
7 www.pinnipeds.org/sealne99.htm
8. "A seal in every port"; *BBC News* 31/1/03 www.news.bbc.co.uk/2/hi/science/nature/2713767.stm
9 "Biology Of Seals of the North East Atlantic in Relation to Seismic Surveys. Thompson, Callan, Duck and McConnell, Sea Mammal Research Unit. SMRU.
10 "Survey Reveals Sea Mammal Numbers"; CBBC Newsround 14/4/04
 www.news.bbc.co.uk/cbbcnews/hi/animals/newsid-3626000/3626269.stm
11 "Sika Habits"; www.sikadeer.com
12 "Of Moose and Men"; *Animals & Men* (see above), Lars Thomas.Issue 10,p 27-28.
13"The Monster of Loch Broom"; *Aberdeen Evening Express* 19/5/1958, Dundee Courier, 16/12/75. All information supplied through personal correspondence from Mrs Wilma Greenwald.
14 Last Refuge of the Monk Seal, Ackerman, D., photographs by B. Curtsinger. *National Geographic Magazine*, Jan. 1992, p 128.
15 Personal email.

Chapter 4.

1. The Dobhar- Chu; Gary Cunningham, *CFZ Yearbook, 2002*. CFZ Publications.
2.Giant Squid, Mystery Boar and Pregnant Snake- Three Irish Animal Stories; Richard Muirhead. *Animals & Men* Issue 14.
3. The Dragons of Scandinavia; Richard Freeman. *Animals & Men* Issue 30.
4. Norwegian Sea Serpents; mjoesomen.no/norwegianseaserpents.htm
5. "Loch Ness Monster could really be a lost Baltic Sturgeon"; Associated Press, 1/2/1994.
6. Seals in Loch Ness; Gordon R. Williamson, taken fro lochnessinvestigation.org/

SILN.html
7. "Andre nets seal of approval". bbc.co.uk/go/pr/fr-/2/hi/uk news/Scotland/2964145. stm
9. "Barriers erected at dam to stop salmon-eating sea lions"; Fox12, Oregon, 1/6/05. kptv.com and KOIN 6 News
10. "Lewis River Incident shoes Sea Lion Problem Persists" *Katu News*, Katu.com

Chapter 5.

1. "Antarctic Scientist Dies In Seal Attack"; BBC News 24/7/ news.bbc.co.uk/1/hi/sci/tech/3090475.stm
2. Beringsea.com; (information on King Island).
3. Sisiutl; www.natureonline.com/legends.html sisiutl 3/03 also wolfsource.org folklore.htm, Palraiyuk.
4. Personal correspondence R. Drozda; NPT (info@nunivak.org), personal email.
5. "Thames swimmer attacked by seal"; *This is London*, 15/8/03thisislondon.co.uk/news/articles/6244848?Source=Evening%20 Standard
6." Surfer bitten by sea lion near Manhattan Beach Pier". 19/6/2005 daily breeze.com
7.KCLU-FM , http://www.kclu.com
8.NBC4
9."Seal bites off woman's nose" ,UnderwaterTimes.com
10. See reference 3.

Chapter 6.

1. "A fishing fleet sees monster", unknown ,"Fishermen badly hit" ?10/11/37; unknown,. Information provided by E.Pinder, Archive Supervisor, Filey District Council via email correspondence.
2. "The Dragons of Yorkshire"; R.Freeman, Animals and Men Issue 14 p 14-19.
3.Animal Tracks: Harbour Seal; Kim. A. Cabnera 1/1/01, beartracker.com/harborseal.htm and related sites.

Chapter 7.

1.Sea Serpents, Seals and Coelacanths ; Darren Naish, *Fortean Studies Vol.7.*John Brown publishing 2001. Also personal web blog.

Chapter 8.

1 " Tree lions in safari park stir"; BBC News 19/10/2004

Chapter 10.

1. *Rocky habitats and Sea Caves of the United Kingdom*; JNCC, jncc.gov.uk/
   publicationajn…/habitat comparison.asp?
2. *Common Seal Distribution in Great Britain and Ire*land; NBN Gateway, searchnbn.
   net
3. Sea Lion Caves; sealioncaves.com
4. Monachus Profiles   The Caribbean Monk Seal www.monachus.org/profiles/
   cariseal.htm

Appendix 1.

1. The Great Sea Serpent, illus. C.M.Park. Dunn, M. *Pall Mall Magazine*,
   London,7/10/1901 p171-180.

BIBLIOGRAPHY

THE SEALS

Berta, A. Sumich, J. *Marine Mammals, Evolutionary Biology.* Academic Press, San Diego, London 1999.

Cleave, A. *Seals and Sea Lions of the World (A Portrait of the Animal World)* Todtri New York 1995.

Godwin, S. *Seals. A Complete Photographic Study.* Headline, London 1990.

Hoelzel ,R. *Marine Mammal Biology, An Evolutionary Approach.* Blackwell Publishing, 2002.

Martin, R.M. *Mammals of the Seas.* B.T. Batsford ,London 1970
.Reeves,R. Stewart,
B.S. Leatherwood, S. *The Sierra Club Handbook of Seals and Sirenians.* Sierra Club Handbooks San Francisco 1992.

Riedman, M. *The Pinnipeds,* (Seals, Sea Lions and Walruses.)University of California Press, Oxford 1990.

Westcott,S. *The Grey Seals of the West Country.* Cornwall Wildlife Trust 1997.

THE LONG NECKS

Binns, R. *The Loch Ness Monster Mystery Solved.* Star, 1988.

Bright, M. *There Are Giants in the Sea*; Guild Publishing, London 1989.

Costello,P. *In Search of Lake Monsters.* Garnstone Press/Panther 1974.

Harrison,P. *The Encyclopaedia of the Loch Ness Monster* Robert Hale, London 1999.
Sea Serpents and Lake Monsters of the British Isles Robert Hale, London 2001.

Healy,T.
Cropper, P. *Out of the Shadows.(Mystery Animals of Australia).* Iron Dark 1994.

Heuvelmans, B. *In The Wake of The Sea Serpents.* Rupert Hart Davis (abridged),London 1968.

Holden, R. *Bunyips: Australia's Culture of Fear* National Library of Australia, Canberra 2001

Kirk, J. *In The Domain of The Lake Monsters.* Key Porter Books, Canada 1998.

LeBlond, P.
Bousfield, E. *Cadborosaurus, Survivor from Deep.* Horsdal and Schubart, Canada 1995.

McEwan, G. *Sea Serpents, Sailors and Skeptics.* Routledge and Keegan Paul, London 1978.
Mystery Animals of Britain and Ireland
Robert Hale, London 1986.

.Mackal, R. *Searching for Hidden Animals(An Inquiry into Zoological Mysteries)*, Cadogan Books, London 1983.

Meurger,M.
Gagnon,C. *Lake Monster Traditions (A Cross Cultural Analysis)* Fortean Tomes, London 1988.

Rife, P. *Americas Loch Ness Monsters* Writers Club Press 2000.

Shuker,K. *In Search of Prehistoric Survivors.* Blandsford, London 1995.
The Lost Ark. Harper Collins, London 1993.

Smith, M. *Bunyips and Bigfoots. (In Search of Australias Mystery Animals).* Millenium Books, 1996.

Witchell, N. *The Loch Ness Story,* BCA 1979.

CFZ YEARBOOK 2007

Picture References

Fig. 1 Cleave, A. *Seals and Sea Lions of the World (A Portrait of the Animal World)* Todtri New York 1995. Leopard Seal, Stefan Lundgren. p54
Fig. 2 Megophias based on Ouedemans (my version) from "*In The Wake of The Sea Serpents*" (IWSS)
Fig. 3 Merhorse after Heuvelmans, IWSS
Fig.4 Long Neck after Heuvelmans, IWSS
Fig.5 Costellos Seal from "*In Search of Lake Monsters*".
Fig. 6 Paperback Cover of Costello's book. Illustration by Brian Froud. Panther edition 1975.Granada publishing; Frogmore, St Albans, Herts AL2 2NF
Fig. 7 Enaliarctos life reconstruction. Berta, A. Sumich, J. *Marine Mammals, Evolutionary Biology.*Academic Press, San Diego, London 1999. p32
Fig 8 Crabeater Seal, M Cameron, National Marine Mammals Laboratory (NMML) website.
nmml.afsc.noaa.gov/gallery/pinnipeds.htm
Fig .9 Steller's Sea lion Group, R. Ream NMML website.
Fig.10 Elephant seal taken from travelmaniac.com website by Karen French.
Fig. 11 a, and b, Mr Picard's creature, taken from A+M.
Fig.12 Mr Aitken's sketch, provided by Mrs W. Greenwald.
Fig. 13 My version of Dungeness Spit Sea Serpent taken from; LeBlond, P. *Cadborosaurus, Survivor from Deep.* Bousfield, E. Horsdal and Schubart, Canada 1995.p2.
Fig.14 My version of Devils Churn Sea Serpent, from Cadbosaurus, p36. As above.
Fig. 15 Hawaiian Monk Seal Underwater; "Last Refuge of the Monk Seal", Ackerman, D ,photographs by B. Curtsinger. *National Geographic Magazine* Jan. 1992, p 128.
Fig. 16 Peer Grove's sketch, taken from IWSS.
Fig 17 Rostron's sketch, taken from IWSS.
Fig. 18 My version of Arthur Grants original sketch, taken from Costello'sbook, p46.
Fig. 19 Kayak decorated with Palraiyuk; Meurger,M *Lake Monster Traditions (A Cross Cultural Analysis).* Gagnon,C. Fortean Tomes, London 1988. p174 (Fortean Times picture Library)
Fig. 20 Decorative pipe with depiction of Palraiyuk, "Ritual Pipe"; Hurst Gallery Archive hurstgallery.com/exhibit/exhibit/past/arctic/historic.htm
Fig. 21 Steller's sea lion at the Brisons, by S. Westcott, *The Grey Seals of the West Country*. Cornwall Wildlife Trust 1997.
Fig 22 a and b, My impression of seal tracks taken from several websites.
Fig.23 My version of Discovery Island Sketch, taken from Cadbosaurus,p 39.
Fig. 24 Batchelor's cute sketch, taken from IWSS.
Fig. 25 My version of Mackintosh Bell sketch, taken from IWSS.
Fig. 26 Life reconstruction of *Acrophoca longirostris* by Darren Naish, taken from

Fortean Studies Vol 7.

Fig. 27 Steller's Sea Lions releasing air underwater, R.Ream NNML, 3/2002.

Fig. 28 Homer the elephant seal, taken from *Grisborne Herald* website, various articles, grisborneherald.co.nz; grisborneherald.co.nz/features/homer.htm.

Fig. 29 Scientist running from elephant seal, taken from *Mammals of the Seas* by I. Collinge, p178

Fig.30 Mediterranean Monk Seal taken from amphoradiving.com/mainsite/seals.html

Fig.31 Northern bull fur seal. Rolph Ream 1992. (NMML)

Fig. 32 Elephant seal taken from Travelmaniacs website, by Karen French.

Fig. 33 As above.

Fig. 34 Crabeater seals, Dr J. Bergston, taken from NMML.

Fig. 35 Stellers sea lion, R. Ream, taken from NMML.

Fig.36 Hooded Seal at various stages of inflation, taken from *Mammals of the Seas,* by F. Breummer, p142-143.

Fig. 37 Angry mother seal, taken from National Oceanic and Atmospheric Administration (NOAA).photolib.noaa.gov/animals/seals.html

Fig. 38 Steller's Sea lions diving, K.W. Kenyon, taken from *Mammals of the Seas*, p111.

Fig. 39 Underwater sea caves of British Isles, taken from JNCC website.Rocky habitats and Sea Caves of the United Kingdom; JNCC, jncc.gov.uk/publicationajn.../ habitat comparison.asp?

Fig. 40 The Fly Sea Serpent, taken from IWSS.

ACKNOWLEDGEMENTS

My thanks and my love go to my wife, Fin, who still doesn't know what I'm going on about, my mum, dad and sister for turning a blind eye to the weird and wonderful array of books that I accumulated in my child-hood, and to Tim, Paul and Emma, the Strange clan, for listening attentively to my babble.
Thanks also to Grimsby Dave F., for his unhealthy scepticism.

Of course thanks must also go to Peter Costello and Bernard Heuvelmans for such a fascinating and charm-ing notion and the slightly obsessional way in which I have come to view the subject over many years.

THE CENTRE FOR FORTEAN ZOOLOGY

So, what is the Centre for Fortean Zoology?

We are a non profit-making organisation founded in 1992 with the aim of being a clearing house for information and coordinating research into mystery animals around the world. We also study out of place animals, rare and aberrant animal behaviour, and Zooform Phenomena; – little-understood "things" that appear to be animals, but which are in fact nothing of the sort, and not even alive (at least in the way we understand the term).

Why should I join the Centre for Fortean Zoology?

Not only are we the biggest organisation of our type in the world but - or so we like to think - we are the best. We are certainly the only truly global Cryptozoological research organisation, and we carry out our investigations using a strictly scientific set of guidelines. We are expanding all the time and looking to recruit new members to help us in our research into mysterious animals and strange creatures across the globe. Why should you join us? Because, if you are genuinely interested in trying to solve the last great mysteries of Mother Nature, there is nobody better than us with whom to do it.

What do I get if I join the Centre for Fortean Zoology?

For £12 a year, you get a four-issue subscription to our journal *Animals & Men*. Each issue contains 60 pages packed with news, articles, letters, research papers, field reports, and even a gossip column! The magazine is A5 in format with a full colour cover. You also have access to one of the world's largest collections of resource material dealing with cryptozoology and allied disciplines, and people from the CFZ membership regularly take part in fieldwork and expeditions around the world.

How is the Centre for Fortean Zoology organized?

The CFZ is managed by a three-man board of trustees, with a non-profit making trust registered with HM Government Stamp Office. The board of trustees is supported by a Permanent Directorate of full and part-time staff, and advised by a Consultancy Board of specialists - many of whom who are world-renowned experts in their particular field.   We have regional representatives across the UK, the USA, and many other parts of the world, and are affiliated with other organisations whose aims and protocols mirror our own.

I am new to the subject, and although I am interested I have little practical knowledge. I don't want to feel out of my depth. What should I do?

Don't worry. We were *all* beginners once. You'll find that the people at the CFZ are friendly and approachable. We have a thriving forum on the website which is the hub of an ever-growing electronic community. You will soon find your feet. Many members of the CFZ Permanent Directorate started off as ordinary members, and now work full time chasing monsters around the world.

I have an idea for a project which isn't on your website. What do I do?

Write to us, e-mail us, or telephone us. The list of future projects on the website is not exhaustive. If you have a good idea for an investigation, please tell us. We may well be able to help.

How do I go on an expedition?

We are always looking for volunteers to join us. If you see a project that interests you, do not hesitate to get in touch with us. Under certain circumstances we can help provide funding for your trip. If you look on the future projects section of the website, you can see some of the projects that we have pencilled in for the next few years.

In 2003 and 2004 we sent three-man expeditions to Sumatra looking for Orang-Pendek - a semi legendary bipedal ape. The same three went to Mongolia in 2005. All three members started off merely subscribers to the CFZ magazine.

Next time it could be you!

Project Kerinci, Sumatra - 2003
In search of the bipedal ape Orang Pendek

How is the Centre for Fortean Zoology funded?

We have no magic sources of income. All our funds come from donations, membership fees, works that we do for TV, radio or magazines, and sales of our publications and merchandise. We are always looking for corporate sponsorship, and other sources of revenue. If you have any ideas for fund-raising please let us know. However, unlike other cryptozoological organisations in the past, we do not live in an intellectual ivory tower. We are not afraid to get our hands dirty, and furthermore we are not one of those organisations where the membership have to raise money so that a privileged few can go on expensive foreign trips. Our research teams both in the UK and abroad, consist of a mixture of experienced and inexperienced personnel. We are truly a community, and work on the premise that the benefits of CFZ membership are open to all.

What do you do with the data you gather from your investigations and expeditions?

Reports of our investigations are published on our website as soon as they are available. Preliminary reports are posted within days of the project finishing.

Each year we publish a 200 page yearbook containing research papers and expedition reports too long to be printed in the journal. We freely circulate our information to anybody who asks for it.

No. Each year since 2000 we have held our annual convention - the *Weird Weekend* - in Exeter. It is three days of lectures, workshops, and excursions. But most importantly it is a chance for members of the CFZ to meet each other, and to talk with the members of the permanent directorate in a relaxed and informal setting and preferably with a pint of beer in one hand. Starting this year-18-20 August 2006 - the *Weird Weekend* will be bigger and better and held in the idyllic rural location of Woolsery in North Devon.

We are hoping to start up some regional groups in both the UK and the US which will have regular meetings, work together on research projects, and maybe have a mini convention of their own.

Since relocating to North Devon in 2005 we have become ever more closely involved with other community organisations, and we hope that this trend will continue. We also work closely with Police Forces across the UK as consultants for animal mutilation cases, and during 2006 we intend to forge closer links with the coastguard and other community services. We want to work closely with those who regularly travel into the Bristol Channel, so that if the recent trend of exotic animal visitors to our coastal waters continues, we can be out there as soon as possible.

Plans are also afoot to found a Visitor's Centre in rural North Devon. This will not be open to the general public, but will provide a museum, a library and an educational resource for our members (currently over 400) across the globe. We are also planning a youth organisation which will involve children and young people in our activities.

Apart from having been the only Fortean Zoological organisation in the world to have consistently published material on all aspects of the subject for over a decade, we have achieved the following concrete results:

- *Disproved the myth relating to the headless so-called sea-serpent carcass of Durgan beach in Cornwall 1975*

- *Disproved the story of the 1988 puma skull of Lustleigh Cleave*

- *Carried out the only in-depth research ever done into mythos of the Cornish Owlman*

- *Made the first records of a tropical species of lamprey*

- *Made the first records of a luminous cave gnat larva in Thailand.*

- *Discovered a possible new species of British mammal - The Beech Marten.*

- *In 1994-6 carried out the first archival fortean zoological survey of Hong Kong.*

In the year 2000, CFZ theories where confirmed when an entirely new species of lizard was found resident in Britain.

Other books available from
CFZ PRESS

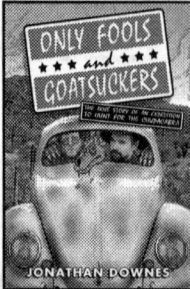

ONLY FOOLS AND GOATSUCKERS
Jonathan Downes - ISBN 0-9512872-3-0

£12.50

In January and February 1998 Jonathan Downes and Graham Inglis of the Centre for Fortean Zoology spent three and a half weeks in Puerto Rico, Mexico and Florida, accompanied by a film crew from UK Channel 4 TV. Their aim was to make a documentary about the terrifying chupacabra - a vampiric creature that exists somewhere in the grey area between folklore and reality. This remarkable book tells the gripping, sometimes scary, and often hilariously funny story of how the boys from the CFZ did their best to subvert the medium of contemporary TV documentary making and actually do their job.

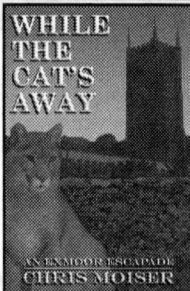

WHILE THE CAT'S AWAY
Chris Moiser - ISBN: 0-9512872-1-4

£7.99

Over the past thirty years or so there have been numerous sightings of large exotic cats, including black leopards, pumas and lynx, in the South West of England. Former Rhodesian soldier Sam McCall moved to North Devon and became a farmer and pub owner when Rhodesia became Zimbabwe in 1980. Over the years despite many of his pub regulars having seen the "Beast of Exmoor" Sam wasn't at all sure that it existed. Then a series of happenings made him change his mind. Chris Moiser—a zoologist—is well known for his research into the mystery cats of the westcountry. This is his first novel.

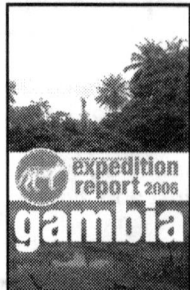

CFZ EXPEDITION REPORT 2006 - GAMBIA
ISBN 1905723032

£12.50

In July 2006, The J.T.Downes memorial Gambia Expedition - a six-person team - Chris Moiser, Richard Freeman, Chris Clarke, Oll Lewis, Lisa Dowley and Suzi Marsh went to the Gambia, West Africa. They went in search of a dragon-like creature, known to the natives as `Ninki Nanka`, which has terrorized the tiny African state for generations, and has reportedly killed people as recently as the 1990s. They also went to dig up part of a beach where an amateur naturalist claims to have buried the carcass of a mysterious fifteen foot sea monster named 'Gambo', and they sought to find the Armitage's Skink (Chalcides armitagei) - a tiny lizard first described in 1922 and only rediscovered in 1989. Here, for the first time, is their story.... With a forward by Dr. Karl Shuker and introduction by Jonathan Downes.

BIG CATS IN BRITAIN YEARBOOK 2006
Edited by Mark Fraser - ISBN 978-1905723-01-0

£10.00

Big cats are said to roam the British Isles and Ireland even now as you are sitting and reading this. People from all walks of life encounter these mysterious felines on a daily basis in every nook and cranny of these two countries. Most are jet-black, some are white, some are brown, in fact big cats of every description and colour are seen by some unsuspecting person while on his or her daily business. 'Big Cats in Britain' are the largest and most active group in the British Isles and Ireland This is their first book. It contains a run-down of every known big cat sighting in the UK during 2005, together with essays by various luminaries of the British big cat research community which place the phenomenon into scientific, cultural, and historical perspective.

CFZ PRESS, MYRTLE COTTAGE, WOOLFARDISWORTHY BIDEFORD, NORTH DEVON, EX39 5QR
w w w . c f z . o r g . u k

Other books available from
CFZ PRESS

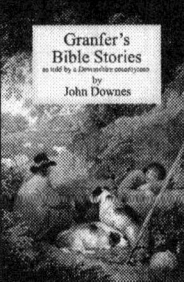